I0135389

Designing the X

Shaping an Unknown Future

Designing the X

Shaping an Unknown Future

Dennis Frenchman, Svafa Grönfeldt, Sigurdur Thorsteinsson

Massachusetts Institute of Technology SA+P Press
Cambridge, Massachusetts

© 2025 Massachusetts Institute of Technology

Designing the X: Shaping an unknown future
Dennis Frenchman, Svafa Grönfeldt, and Sigurdur Thorsteinsson

Project Lead
María Esteban Casañas

Editors
Matilda Bathurst, development editor
Elizabeth Hamblin, copy editor

Design
Sigrún Sæmundsen, Igor Micevic, Tiziana Alocci

Publisher
SA+P Press
MIT School of Architecture and Planning
77 Massachusetts Avenue
Cambridge, MA 02139

Distributed by The MIT Press
Printed and bound in Italy by Fontegrafica Lab.

ISBN: 9780998117089

Library of Congress Cataloging-in-Publication Data applied for.

All rights reserved. No part of this book, except as noted below, may be reproduced in any form by any electronic or mechanical means, including information storage and retrieval systems, without written permission from the publisher.

Every reasonable attempt has been made to identify the owners of copyright. Errors or omissions will be corrected in subsequent editions.

Table of Contents

Foreword   8

Acknowledgments   14

1   Designing the X   17

2   Design in Time   43

3   Progettare   63

4   Flow with Complexity   89

5   The Collective and the Conductor   127

6   The Path Forward   147

Epilogue   169

Appendix   173

Cases

Frolic Community — Building Equitable Communities   41
Biobot — Improving Global Health   61
Quipu — Embracing Informality   87
Roofscapes — Reducing the Heat Island Effect   117
Blue Lagoon — Playing Chess with Nature   118
Learning Beautiful — Preparing for a Digital Future   145
Bandhu — Reforming Urban Migration   167

To the innovators who dare to design.

Foreword
by Mauro Porcini
President & Chief Designer Officer | Samsung

Design is not a job. It is not a tool or a strategy. Design is a way of being in the world.

It is how we choose to see — with empathy.
How we choose to listen — with curiosity.
How we choose to act — with imagination, courage, and care.

When I first encountered the ideas behind Designing the X, I didn't just read them. I recognized them. They resonated deeply, not because they were familiar, but because they revealed something I had always felt.

This book speaks to the designer in all of us — not only the one who sits in front of a screen sketching icons or products, but anyone who wakes up each morning and tries to make something better. A conversation. A company. A society. A culture. A future.

Design is not just a discipline. It is a way of seeing — and of feeling — the world. It is how we respond to complexity not with fear, but with an open mind. It is how we move from reaction to intention, from function to meaning, from isolation to co-creation.

Designing the X was conceived in that fragile and fertile space between ambiguity and action. It emerged from conversations across continents and disciplines, among designers, educators, scientists, entrepreneurs, and artists who shared a single belief: that design has the power not just to solve problems, but to reveal them; not just to improve systems, but to reimagine them; not just to answer questions, but to ask better one.

The "X" in the title is deliberate. It represents the unknown — the emergent, the unformed, the space of possibility. It is the variable that can only be defined through experience, through process, and through synthesis.

That "X" is both a mystery and a promise. It stands for the space where our intuition begins. It is the blank canvas, the leap of faith, the trembling silence before the music starts. And it is in that space that the beauty of design — and the truth of humanity — comes alive.

I've spent my life exploring this mystery. From a small town in Italy, drawing with passion in the margins of notebooks, to boardrooms in St. Paul, New York and Seoul, leading global teams across brands and borders — I've seen firsthand what design can do when it's guided not by ego, but by purpose. When it becomes a dialogue, not a monologue. When it welcomes diversity and contradiction and still finds harmony.

Designing the X captures this beautifully. In that sense, this book is not a map. It is a compass. It doesn't offer a single path forward — it offers a way to navigate the future.

We live in an era of supercomplexity — where the pace of change is accelerating beyond the reach of any single method, model, or mindset. Traditional disciplines, even when combined, often fall short in addressing the entangled crises of our time: environmental collapse, social fragmentation, technological disruption, and economic inequality. In the face of these challenges, design offers something essential: a means to think and act in conditions where cause and effect are opaque, where the outcome is uncertain, and where the stakeholders are many.

Design invites us to engage the full spectrum of human experience — rational and emotional, analytical and intuitive, scientific and poetic. It demands that we hold paradoxes without resolving them too quickly. It asks us to remain open, adaptive, and humble. And it insists that we do so together.

The future cannot be designed in isolation. It must be co-created.

This is the deeper argument of *Designing the X*. Drawing from decades of work across industries, cultures, and academic institutions, the authors propose a vision of design not as a set of tools, but as a mindset: a human capacity for synthesis, meaning-making, and transformation.

In this vision, the designer is not a solitary genius. The designer is a conductor — a guide, a translator, a connector of humans and ideas. They are "people in love with people", as I like to call them. With care and sensitivity, the design conductor orchestrates the voices of many: users, stakeholders, scientists, business leaders, artists, policymakers, citizens. They do not impose harmony; they make space for discord and diversity to coexist productively. They give form to dialogue. They turn friction into fuel.

The process of design, as this book reveals, is not linear. It is fluid, recursive, embodied. Like a vortex, it gathers insight, energy, and tension as it moves. At its best, it is both grounded and visionary — capable of engaging reality as it is while imagining what it might become.

Designing the X takes us on a journey through this process. It offers frameworks grounded in practice and illustrated through case studies that span the globe — from urban innovation and affordable housing to sustainable development and speculative futures. It introduces us to polymathic teams and radical thinkers. And it reminds us that design is not about perfection. It is about progress. It is about participating in the ongoing evolution of the world around us.

What makes this book so timely — and timeless — is that it does not celebrate design as an isolated act of creativity. Instead, it positions design as an integrative force, one that can bridge the artificial boundaries between disciplines, between people, and between present and future.

Design is not just a response to crisis. It is a rehearsal for what comes next.

In that spirit, this book invites us to reframe design as a form of hope — active, critical, and courageous. It is hope not as wishful thinking, but as intentional action in the face of uncertainty. Hope as strategy. Hope as infrastructure. Hope as method.

There is, of course, an urgency to this invitation. We are living through a moment of profound transition — cultural, technological, ecological. The stakes are high. But the authors of *Designing the X* do not succumb to despair. Nor do they fall into naïve optimism. They choose what I would call "creative optimism" - a belief that we can and must imagine alternatives, not in spite of complexity, but because of it.

They remind us that design is not neutral. It always reflects choices —
about who is included, what is valued, and what future we are building
toward. Design can be extractive or regenerative, exclusionary or
inclusive. Which is why ethics, empathy, and equity must be at the core
of any design process.

The book makes a strong case for design literacy — for bringing design
into education, into leadership, into systems thinking. Not because
everyone should be a designer in the traditional sense, but because
everyone is already shaping the world around them in ways big and
small. If we are all participants in shaping the future, then we all need the
tools and mindsets to do so wisely.

Above all, *Designing the X* is a celebration of plurality. It honors
complexity without collapsing it. It respects expertise without fetishizing
it. And it elevates the lived experience of people — users, communities,
collaborators — as vital sources of insight and innovation.

This is a deeply generous book. It gives us language to describe what
many of us have long felt: that design is not a luxury, not an afterthought,
not the "last step" in innovation. It is the connective tissue — the means
by which ideas become actions, and actions become impact.

It is a book that dares to ask big questions, while offering practical
pathways forward. It does not offer quick fixes or silver bullets. Instead, it
gives us a richer, deeper understanding of what it means to design in an
age of complexity, change, and interconnectedness.

To read this book is to be reminded of something essential: that we are all designers, whether we call ourselves that or not. That every choice we make — what to build, what to change, what to protect — is a design decision. And that the future is not written. It is designed.

To the quiet leaders and bold visionaries,
To the strategists, artists, engineers, and activists,
To the rebels who lead with love,
To the teams who argue, laugh, and create together,
To the teens drawing spaceships on their notebook,
To my daughter and son, and their generation, who already design their days with wonder —
This book is for you.

And if you, like me, believe that ideas are seeds and design is how we help them grow —then welcome. You are exactly where you need to be.

Mauro Porcini, an Italian designer and innovation leader, is the author of *The Human Side of Innovation: The Power of People in Love with People*. Prior to joining Samsung, he served as the first Chief Design Officer at 3M and at PepsiCo. He has received multiple awards, and his work has been featured in numerous books and articles on design and innovation around the world. More information at www.mauro-porcini.com.

Acknowledgments

Contributors to *Designing the X* and the associated research project extend far beyond the authors, who wish to gratefully thank the following individuals for their insights, inspiration, and hard work:

María Esteban Casañas, MITdesignX project lead, for her resourcefulness, resilience, and meaningful contributions to virtually every aspect of the project.

Gianandrea Giacoma, psychologist, psychotherapist, and consultant (Design Group Italia), for his incredible insights and ability to enrich and expand our thinking throughout this journey, as well as our dear colleague Francesco Zurlo, dean of the School of Design of the Politecnico di Milano, for his substantive input throughout the process and writing in Chapter 2. We are also deeply grateful to Dale Dillavou, science advisor, for his thoughtful guidance and generous support.

The editorial team: Matilda Bathurst, our outstanding development editor, was central to this work. Her intellectual rigor, creativity, and editorial insight helped us clarify and articulate our ideas. Elizabeth Hamblin provided brilliant copy editing and support. Our sincere thanks go to Melissa Vaughn, director of communications at MIT School of Architecture and Planning, and to SA+P Press for providing the direction and support needed to make this publication possible. These tireless comrades-in-arms kept us enthusiastic and on track — thank you for your unwavering support and belief in this project.

The design team: Sigrún Sæmundsen (book design) and Igor Micevic (graphic design) brought the concepts to life with care and creativity. Their thoughtful work shaped not only how the book looks, but how it feels and communicates. We're deeply grateful for their contributions.

We thank the following individuals for their dedicated work during the interview phase of this project and their generous contributions: Gilad Rosenzweig, executive director, MITdesignX; Michael Stradley and Smriti Bhaya, research affiliates; and from Politecnico di Milano: Carla Sedini, assistant professor, Department of Design, with the collaboration of Alessia Orizio and Sara Prevosti; AXIS Inc. and Yasuyuki Hayama, now assistant professor at Kyushu University in Japan. Their contribution throughout the interviewing process was invaluable.

We are deeply grateful to our students and their MITdesignX teams for their inspiring courage and dedication in transforming bold ideas into purposeful action — we learn from you every day. We also thank the Blue Lagoon management team for their thoughtful collaboration throughout this journey.

Finally, we owe a debt of gratitude to each of the 67 interviewees (see the Appendix), who generously shared their time, deep experience, and knowledge, informing and enriching this book. They are a living testament to the fact that these ideas are alive and thriving in the real world. Thank you for the opportunity to synthesize them in *Designing the X*.

1 Designing the X

No model or mathematical formula alone can capture the complexity of our world with all its emotional, cultural, and human variables that are impossible to measure. Hence, we must design.

We had gathered in Iceland to develop the ideas for *Designing the X*: a group of strategic designers, practicing across fields including city planning, entrepreneurship, education, management, and public health, linked by a shared intuition that design held the key to engaging the challenges of our time.

All around us, we saw those challenges played out at scale — from economic disparity and mass migration to rising sea levels and extreme weather events. We had come together to connect the dots between our different areas of practice: to look deeper into the design process by investigating how design acts as a vehicle for innovation and taps into the full scope of our human capabilities.

After months of research and an intensive week of discussion, we were ready to take a break. On the invitation of our co-author Sigurdur Thorsteinsson, we traveled to the Blue Lagoon, the globally renowned natural hot spring located on the Reykjanes Peninsula.

Heated by turbulent forces deep underground and harnessed to generate clean energy before entering the lagoon, the milky-blue waters are recognized for their healing properties. Set within an ink-black landscape of ancient lava, underwater the rock is coated with a white, porcelain-like finish of minerals. It's mesmerizing, conjuring associations of an otherworldly paradise — all too easy to forget that we are submerged within a volatile living landscape.

"Design has the capacity to synthesize across different domains, weaving together complex information, bold ideas, inspirations, and technologies — to craft novel solutions."

Andrea Chegut
Innovator, research scientist, and financial econometrician
Interview **16**

As we drifted calmly in the azure waters, the sky suddenly blazed red, reflecting the flames of a nearby eruption. Beyond the lagoon's protective walls, glowing yellow-orange lava flowed freely on its way to the sea, threatening to obstruct the only route of escape.

Wondering whether it was time to panic, we turned to Sigurdur for reassurance. He responded with bemused Icelandic calm:

"We are playing chess with nature."

Since that day in midsummer, we've come to view the Blue Lagoon as the perfect expression of contemporary conditions of life on Earth. Enmeshed within the play of nature, there is no retreat or escape. Each day, we wake up to a world that's shifting beneath our feet, where the ground feels less stable, where the rules are constantly changing.

Each one of us is living in a situation where the old certainties have crumbled, replaced by a web of complex interconnections that cannot be fully comprehended. The climate and the economy — the very systems our species depends upon — are intertwined and impossible to untangle. We're left to navigate an unpredictable terrain where the stakes are higher and the outcomes more uncertain than ever before.

The Challenge of Supercomplexity

"In the face of super-
complexity, we must
resist the urge to
reduce one perspective
to another. Instead,
we should cultivate
dialogue between
distinct points of
view — A and B — not
to merge them into
sameness, but to let
their interplay generate
deeper understanding."

Luisa Damiano
Epistemologist of complex
systems philosophy of emerging
sciences and technologies
Interview **23**

How and why did such "supercomplex" systems emerge? They are surely not a product of human intention. Few people could envision living and working, let alone swimming, amidst a volcanic eruption. Yet for Icelanders, volcanoes have become part of life, just as they were for the early Viking settlers. In Norse mythology, eruptions signaled the wrath of Surtr — a giant who summoned rivers of fire from deep within the Earth to clear the land of wayward humans (Lindow, 2002). Today, across the world, we are witnessing a comparable unstoppable flow: No one could have foreseen the complex interplay of data, algorithms, and multiple human systems that are reshaping our daily lives in subtle yet increasingly profound ways.

The condition we refer to as "supercomplexity" extends beyond measurement and formulas. It involves dynamic physical, behavioral, economic, and environmental factors in which the logic of cause and effect is not always clear or observable, and the relationships between the parts and the whole are constantly shifting (Barnett, 2000). Supercomplex systems are opaque, leading to problems characterized by:

Impenetrability — where should the problem-solving process begin?

Uncertainty — how do parts of the system relate, or not, in space and time?

Fluid emergences — in which solutions are not bound by predetermined rules, but evolve from within the problem-solving process itself.

Supercomplexity is accelerating in every aspect of life. This is not, as some have argued, merely the perspective of each new generation; our current situation can be likened to the exponential accumulation of human knowledge, estimated by IBM to be doubling at this point every 12 hours (Schilling, 2013). Such acceleration is beautifully visualized in Barrett Lyon's

Opte Project, which seeks to accurately map the growth of the internet in real time. As revealed in Lyon's timelapse video, each complication springs from its predecessor in a never-ending accumulation of change, prompting psychologist Mihaly Csikszentmihalyi (1990) to question: "Is life becoming too complex for survival?"

At this moment of increasing doubt and rising panic, there is a source of equanimity in Sigurdur's observation: "We are playing chess with nature." The condition of supercomplexity should not be unexpected. Increasing complexity has been a function of our universe from its first moment of existence as an explosion of unformed energy through the appearance of atoms, elements, stars, galaxies, planets, living organisms, and the intricacies of human consciousness. Each stage of emergence, instigated by imbalances between a system and its environment, is marked by a leap in complexity, which drives the universe to ever-greater levels of interconnection.

This growing net of relations has generated what we know as life on our planet. However, increasing complexity is also characteristic of what we experience as crisis. As the climate warms and human population continues to consume, the habitable space and resources available to support life on Earth are now critically out of balance. Among many daunting statistics, global wildlife populations have declined nearly 73% on average in

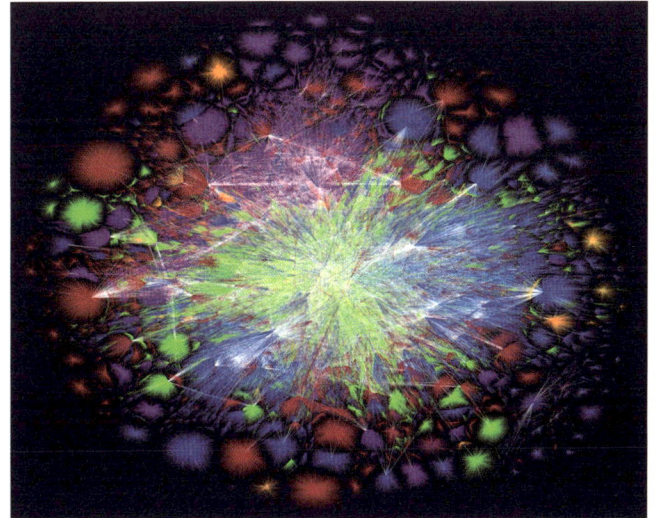

Barrett Lyon / The Opte Project.

just a few decades, with a shocking 95% decline across Latin America (World Wildlife Fund, 2024). In many human populations, we see advancing scarcity of food and water, health crises, displacement, and unchecked urbanization, which pose problems at a scale of unprece‐ dented complexity. For example, the population of Africa is predicted to double by 2050, adding 1 billion people to urban areas (African Development Bank, 2023) — a staggering increase led by the city of Kinshasa in the Democratic Republic of the Congo, which is expected to reach 35 million people in the coming decades (Wahba & Ranarifidy, 2018).

The Why of Design

"True innovation happens
when ideas collide,
inspiring us to see
beyond our own limits."

Kengo Kuma
Architect
Interview **36**

What will it take to regain equilibrium and intentionally engage supercomplexity? What methods are best suited to supporting the balance of life on the planet?

In this book, we argue that conventional approaches to understanding and problem-solving fall short of these goals. Despite the hype around AI and new predictive technologies, the accumulation and analysis of ever more data are not sufficient to tackle the challenges ahead. At a time when the analytic paradigm favored by scientists and technologists is relied upon to solve the world's problems, we propose that design can act to synthesize new approaches, as well as transform and advance the capabilities of many disciplines. Indeed, by providing a crucial missing piece of the puzzle, we believe that design is essential to the quest for a more sustainable and equitable future.

The team behind *Designing the X* has arrived at this shared conviction through diverse routes: from designing large-scale, technology-enabled urban environments, to building successful ventures centered on social impact, to radically rethinking the way we produce and use consumer products. This broad compendium of experience and knowledge has been supplemented by insights coming from physics, neuroscience, anthropology, and philosophy, among other fields, as well as 67 interviews with global thought leaders and practitioners about the role and practice of design.

These dialogues have deepened our understanding of the inner mechanisms and intuitive discoveries of the design process and how that process can be applied to engage supercomplexity. The voices of our interviewees are present throughout the book, appearing as quotations that direct the flow of the text.

As we introduce our distinct argument and set of tools to engage supercomplexity, it is important to note that our thinking also builds upon a strong foundation established by other scholars and design practitioners. Books such as *Design Unbound: Designing for Emergence in a White Water World* by Ann M. Pendleton-Jullian and John Seely Brown (2018), *Design for a Better World* by Don Norman (2025), and *Reimagining Design: Unlocking Strategic Innovation* by Kevin G. Bethune (2024), among others, set a precedent when it comes to articulating a broader scope for design in society.

"How we collect and use data, and how it impacts people's lives, is fundamentally a design question."

Molly Wright Steenson
Historian of Artificial Intelligence
Interview **57**

"Design is not a content area, design is a field of procedures and methods."

Don Norman
Author, Professor Emeritus
Interview **47**

"Design is both a way
of doing and a way of
seeing the world."

Nicholas de Monchaux
Designer, author, educator
Interview **24**

Designing the X presents a new take on the "why," based on a deep investigation of exactly how design acts as a vehicle for innovation. Our argument is rooted in an in-depth exploration of what design is, how it originates, and its role in shaping the human condition. This informs our proposal (in Chapter 6) for where design fits within the currently dominant science, technology, engineering, and mathematics (STEM) disciplines, especially when it comes to education.

By prospecting for ideas, information, and shared patterns of thinking across different areas of knowledge, it became apparent that design is central to contemporary thinking, revealing both the deep origins of the practice and its power to shape the future in sync with other disciplines. We are indebted to a range of sources in addition to our interviews, broadening our outlook to areas far beyond typical design discourse — from theoretical physics and neuroscience to linguistics and philosophy.

In opening a conversation about the positive role of design in remaking the many realities we each experience, we have raised to the surface what is already present in the zeitgeist. This multiplicity is modeled in the book by the interviews, our perspectives from practice and teaching, and ideas and insights from many different angles. Together, these represent the ethos of synthesis, which serves as a guiding theme and organizing principle of the book.

In Pursuit of the X

At its root, design can be understood as a method of synthesis — an innate human ability that relies upon intuition, prediction, and the facts of the present moment to envision an alternative future state and develop pathways to attain it. Design involves inventing new wholes that are more than the sum of their parts, revealing problem-solving opportunities that are not available to analytic methods of reasoning that deal with the parts alone. The difference between the sum of the parts and the greater whole is the "X" in our title.

This process of synthesis can be compared to the reaction between two poisonous elements: sodium (a reactive metal) and chlorine (a corrosive gas). When combined, the two react to produce a new compound: table salt, the mineral of life.

In this case, the transformation from two separate elements to a new and distinct whole can easily be understood in terms of elemental chemistry. However, in design, the apparent "leap" from an assortment of parts to a greater whole — the holistic and contextual re-weaving of conditions so that the problem could not exist — has not yet been investigated in depth.

Too often, the X is credited to the mysterious "magic" of creativity — a mythology that serves to lionize the designer in the public imagination yet belittle their role when it comes to addressing real-world problems. We believe it is time for a closer look "under the hood" of the design process, with the goal of identifying a set of design principles and strategies that can be applied by innovators in any field.

Designing the X requires embracing the uncertainty of entering the unknown; it is only by leaving the security of the known present that we can identify a path forward into an envisioned future. By living in pursuit of the X and applying the know-how of design, we open ourselves to the opportunities latent within supercomplex situations — discoveries made possible by innovating within the fluid medium of the design process itself.

Our approach is indicative of a shift in consciousness occurring across multiple fields of practice. It has become increasingly clear to many, whether philosophers or artists, sociologists or scientists, that something is missing in our current view of reality: an exclusion of the less measurable human factors that shape much of what is perceived, and therefore, what can be conceived as possible (Deacon, 2012, p. 108). It is time to expand the aperture.

To understand the context of this shift, it's worth taking a step back to the 17th century, when the Eurocentric frame of perception became increasingly dominated by analytic thinking, in which beliefs are tested on the basis of physical evidence that can be observed, measured, and applied to form conclusions. These new methods of thinking made way for the scientific discoveries of Copernicus and Newton, and the process of deduction that led Descartes to his famous pronouncement of certainty: cogito, ergo sum, "I think, therefore I am."

Thinking, specifically the capacity for inferential reasoning, was posited as the primary form of knowledge and sole means of identifying facts — in this case, as proof of Descartes' own existence. The mind, dualistically divorced from the body, would henceforth be the primary engine for a new approach to problem-solving: the interrogation of apparent "reality" to decode the verifiable "facts" of how the world functions, breaking down experience into ever more detailed, ever simpler units in the hope of disclosing a fundamental set of causes and effects. By such reasoning, a whole cannot be more than the sum of its parts; human beings are separate from the world they observe, and there is no possibility of accessing the X.

It is important to note that the formulation of the scientific method had a precedent in the work of non-Western scholars and can be traced back to ancient China (Needham, 1995) and the mathematical discoveries that shaped the culture of the Islamic world (Al-Khalili, 2011). However, it was in Europe that analytic thinking became a dogma of sorts, understood as the source of agency for effecting change in the world (Shapin, 2018). With the emergence of the Industrial Revolution, the attempt to decode the secrets of nature and apply those laws for human gain became the environmental endgame. Advancing via multiple technological discoveries through to the present day, this trajectory of apparent progress has led to unprecedented and often beneficial changes in living conditions, but with deep, unforeseen consequences: the destruction of our habitat on Earth.

A Balancing Act

Today, we find ourselves at a crossroads. Social and ecological problems have reached a breaking point. What's more, these problems are too intractable to resolve by conventional means, and the position of analysis as the primary means of truth-seeking and problem-solving is no longer so stable. All it takes is a recalibration of scale to recognize that what we rationalize as "reality" is only a partial placeholder. At the level of subatomic particles, the laws of classical physics no longer apply; this is a realm where cause and effect are seemingly divorced, the future can influence the past, and matter can fundamentally change depending upon how — or if — it is observed (Baclawski, 2018). The method of analytic reasoning has become its own undoing, rationally revealing itself to be unreasonable.

Unreasonable, that is, if analytic reasoning (enshrined by the scientific method) is accepted as the sole arbiter of truth (the "real") at the expense of other approaches to knowledge. The assimilation of quantum mechanics as part of our everyday understanding of the world will inevitably produce vertigo, but it can also be a new way of finding our feet. No longer standing on solid ground, but moving in a medium that is far more interesting and changeable than any framework of truth we might wish to impose upon it.

Perhaps the test of a radical worldview is how quickly it becomes "common sense." Certain distinctions are proving to be insufficient — east and west, black and

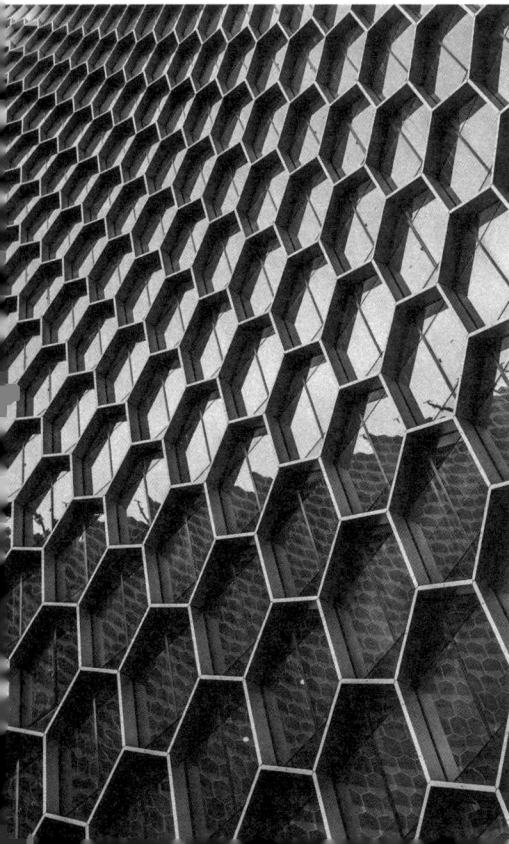

white, male and female, to name just a few of the binaries that no longer fit the world in which we live. Likewise, the integration of methods of analysis and synthesis, of mathematical and human truths, of quantum mechanics and sensed experience, does not require a feat of mental gymnastics. To expand the aperture simply means to open the mind and to see as we are able.

This observation is echoed by Andrea Moro, professor of linguistics at the Institute for Advanced Study in Pavia, Italy:

The opposition between humanism and science was functional to the emancipation of science from theology. But now I think the terms "humanism" and "science" are an obstacle to understanding reality. It seems to me that those terms divide what seems to be more united than ever. There can't be a real distinction between the human and what is real. And so we have a new challenge: to abandon this dichotomy — humanistic and scientific — and reconcile them methodologically, epistemologically. We have to use all tools to understand reality (Eberly, 2022).

Scratch the surface of neurobiology, and it becomes apparent that this balancing act between analysis and synthesis is innate to the human brain. It has long been understood that brain function is divided into two interdependent hemispheres: while the right hemisphere forms a holistic picture of our place in the world by attending to context, connections, intuitions,

"Design is an endless frontier"

John Ochsendorf
Engineer, designer, educator
Interview **48**

31

and emotions, the left hemisphere sees the world as constructed of isolated, unchanging elements and quantifiable facts that can be measured and categorized. The right hemisphere is open to the implicit, while the left focuses purely on the tangible and explicit. Furthermore, while the right hemisphere of the brain sees the synthesis of both modes of sense-making, the left is unaware of the right and is therefore entirely confident of its own analytically driven conclusions (McGilchrist, 2019).

Psychiatrist and neuroscientist Iain McGilchrist argues that human culture has veered toward overvaluing the functions of the left hemisphere, a tendency that has moved in cycles across time and has become increasingly dominant since the scientific revolution in the 17th century. This imbalance has had profound and damaging effects, arguably laying the groundwork for the current ecological crisis. In technologically advanced, urban-centric societies, there is an inclination to view our species as separate from nature and the planet as inanimate matter, reduced to an accretion of resources to extract for our own ends.

Indeed, it is our own end — the demise of our species, among the many extinctions due to ecological disruption — that we risk approaching due to left-brain dependence. Unreceptive to information that is not present and provable, the left hemisphere lacks the capacity to envision possibilities beyond what it can see. It is therefore doomed to repeat the same errors, trapped in a closed box of disembodied abstractions and analytic segmentation (Nielsen et al., 2013).

"Design can go beyond problem-solving to become problem-framing, transforming intractable dilemmas into opportunities for radical reimagining. Artists and designers create spaces where competing values can coexist, enabling communities to rehearse new ways of being together."

Azra Aksamija
Artist and historian
Interview **2**

To confront supercomplex problems, our human challenge is to restore the balance between these distinct but complementary ways of understanding the world: the analytic and instrumental, alongside the synthetic and intuitive. This is where design fits in, the missing puzzle piece that enables us to complete our picture of the world and what is possible within it.

How?

1. Design differs profoundly from analysis in that it accepts human intuition, visions, emotions, and beliefs as equal in value to known facts.

2. Design pushes for an in-depth understanding of the interconnected context in which problems and solutions emerge, requiring the involvement of stakeholders. This differs from analytic reasoning, which attempts to isolate the essence of a problem within a simplified, reductive model.

3. Design deals with synthesizing new systems in the context of an envisioned future, rather than trying to dissect existing systems to understand reality in the present.

These and other relationships assert the potential of design to make way for more powerful methods of predicting and producing outcomes. Rather than seeking to replace the dominant forms of knowledge represented by science and technology, design serves to synthesize new knowledge while opening alternative pathways for innovative problem-solving.

"In Antarctica, I learned to be at ease with change. When the environment is extreme, fast, and risky, the models you've learned no longer work — you have to let them go. You have to change how you change."

Chiara Montanari
Engineer, Antarctic explorer
Interview **43**

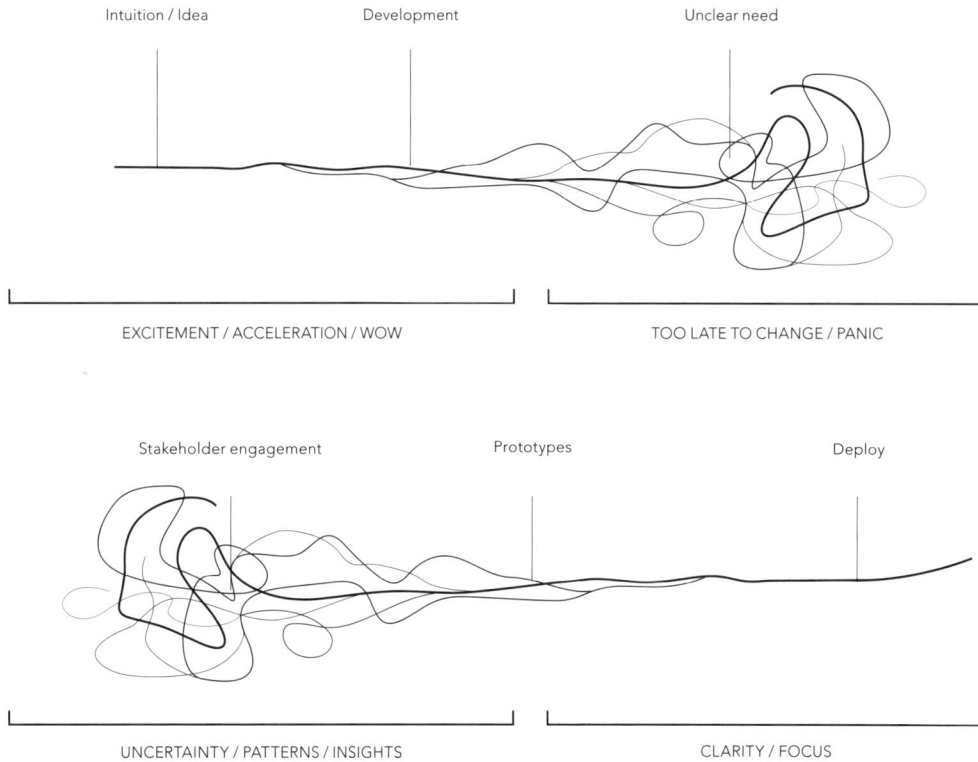

Intuition / Idea   Development   Unclear need

EXCITEMENT / ACCELERATION / WOW   TOO LATE TO CHANGE / PANIC

Stakeholder engagement   Prototypes   Deploy

UNCERTAINTY / PATTERNS / INSIGHTS   CLARITY / FOCUS

Figure 1.1 | Differing Routes on the Innovation Journey. (Top) A solution-based approach applies a preconceived idea from the start, largely ignoring context and stakeholder needs. Without continuous exploration, testing, and a clear understanding of the innovation context, critical flaws emerge too late, leading to costly consequences and failed deployment. (Bottom) In contrast, a design-driven process — rooted in continuous engagement, iteration, and co-creation — leads to relevant and effective solutions by adapting to its contexts and improving through constant feedback to drive scalable impact. A design process engages with complexity up front, using synthesis to select among possible paths forward until a preferred solution(s) emerges.

Finding your flow

Designing the X guides the reader through a series of strategies for innovators to advance their problem-solving capacity — both at the level of envisioning the unprecedented and taking the steps to translate that vision into reality.

The structure of the book has a sequential flow. However, the reader may wish to take any path that captures their interest, even starting at the center and cycling outward.

In **Chapter 2: Design in Time**, we reflect on the history of design as a fundamental human ability and an active force in changing the trajectory of human culture. It is important to recognize that good design is far more than an aesthetic strategy that serves to fuel consumerism. The emotional resonance and sensual appeal that design brings to an everyday object might equally be applied to ideas, events, and causes, transforming human perceptions and introducing new ways of thinking and acting. As such, design plays a "predictive" role in conditioning the possibilities for future innovation. This capacity — intangible as it is powerful — has been overlooked and underestimated. From shaping tools to give them meaning and power over nature, to inventing new forms of sustainable cities, design has a history of producing alternate realities.

Expanding upon a concept borrowed from the Italian legacy of design, **Chapter 3: Progettare** outlines the principles that serve to transform a vision into tangible reality. Contrary to the popular myth of the designer as a maverick "creative" who conjures "pies-in-the-sky" from thin air, the practice of design is grounded and strategic. To plan ("progettare") is always to work in relation to a future that has not yet occurred, and designers are distinguished by a high tolerance for navigating uncertainty. We argue that this is not an elusive superpower but an aptitude developed by having access to certain key strategies for directly addressing complexity and working through the unknown. Informed by recurrent themes revealed in our research, we propose a set of engagement principles that provide effective touchpoints for anyone seeking to address supercomplex problems.

Chapter 4: Flow with Complexity dives deeper into the design process and methodologies, mapping the concurrent cycles that create conditions for the emergence of a design solution. We liken this process to a vortex, a natural phenomenon that emerges when a flow (whether of air, water, or thought) is interrupted. A vortex is not a singular entity but a dynamic event, in which each move determines the next, taking different shapes and directions as the flow is maintained.

Novel approaches emerge from the form of the process itself, and one may enter the vortex at any point or scale of the problem. To arrive at desirable outcomes, we apply certain methods to generate and constrain choices as we move with accelerating speed down the vortex. The objective is not necessarily to optimize a single correct answer. Instead, the vortex feeds the dynamics of emergence; it frames a reality in which desirable outcomes are probable and therefore predictable.

Who are those that navigate the vortex? **Chapter 5: The Collective and the Conductor** describes the orchestration of an inclusive and "polymathic" design process, one that takes care to integrate the multiple different perspectives required for addressing supercomplexity. Led by an experienced "design conductor" who manages and guides the process forward, the design team includes not only professional designers and experts from multiple fields of knowledge, but also stakeholders who can be considered experts in the context. These diverse team members play a crucial role, both as an important source for facts and ideas and as decision makers and sculptors of the process. Today, it has become something of a cliché for designers

to assert an anti-didactic, non-prescriptive approach to problem-solving, reflecting a cultural shift away from the 20th-century "design master" and toward an ideal of ground-up design involving stakeholder engagement. However, the proof is in practice. Our view is not based on the "morality of the moment," but a pragmatic recognition of how consensus solutions are reached.

Chapter 6: The Path Forward explores the implications of incorporating design education across all disciplines, fostering and developing the innate capacity of the mind to synthesize new realities. In the current STEM-oriented education system, the sciences and mathematics have been of primary importance. However, this overemphasis on analytic reasoning limits the toolkit for tackling future challenges, equipping the next generation with only part of what is needed to negotiate an increasingly supercomplex world. In contrast, design engages the whole, just as a whirlpool engages a stream. The process of intertwining analysis with synthesis produces a balanced and context-driven approach to living and acting in this world — ultimately, allowing that world to evolve into the place we envision and aspire to.

The Optimism of Experience

The framework above has been developed in the spirit of "creative optimism." As a group of educators and design practitioners, we believe it is possible to apply the full scope of our capacities as human beings to co-create better futures.

This conviction is grounded in experience. We have seen our insights played out at the level of industry and academia. Throughout the book, we draw on our industry experience and our work at the MITdesignX program at the Massachusetts Institute of Technology (MIT), Design Group Italia (DGI), and insights from our colleagues at the Design School at Politecnico di Milan.

MITdesignX operates in the context of one of the world's leading scientific institutions known for innovation and new thinking. As an experiential learning program founded in the MIT School of Architecture and Planning and based at the MIT Morningside Academy for Design, the program is dedicated to accelerating innovation in design and the human environment, engaging participants working across MIT in disciplines as diverse as bioengineering and art. Working with polymathic teams, many linked to the sciences, we have witnessed how synthesis and original thinking emerge through the design process. Projects ranging from tackling urban migration in India, to curbing opioid addiction in the United States, to developing cities more resilient to climate change, demonstrate the synergetic power of science, technology, and design working together.

Design Group Italia, headquartered in Milan, is one of the largest multidisciplinary design studios in Italy. Their mission is to shape the future through data-driven research and design, creating physical and digital solutions that drive meaningful change.

Politecnico di Milano's School of Design is a leading center of design research in Europe, and its students and faculty have become international figures in the construction of contemporary design culture. The school blends Italian design heritage with innovation, offering programs in product and fashion design as well as digital and interaction design.

"Design is a
collective act."

Roberto Verganti
Scholar and author
Art and innovation,
design theory
Interview **63**

Our networks and professional careers have granted us access to some of the world's foremost thinkers on design, prompting us to curate a series of interviews to enrich our observations and provide alternative points of view. These insightful conversations introduce the informed perspectives of professionals and academics working across an extraordinary spectrum of technical and humanistic endeavors in the United States, Latin America, Europe, Africa, and Asia — ranging from unleashing new sources of renewable energy, to designing environments on Mars, to promoting social equity through technology. Their viewpoints, quoted throughout the book, have inspired both its concept and content.

Learning From Our World

When we reflect upon the conditions of the world in the 21st century, it is astounding to recognize that everything we encounter has been shaped by human intervention (consciously or unconsciously) — in other words, designed. As a fundamental human ability, design is positioned at the nexus of human existence and our environmental context on the planet, bringing together all disciplines and phenomena of human culture. As such, we consider design to be a fundamental way of transforming the world. In fact, that transformation is always occurring, and the power of design is evident everywhere — for better or worse.

Working backward from an envisioned future and forward from the present, within the vortex of design we can choreograph a path between current conditions and an ambitious future outcome, whether or not that outcome is possible in the present. Finding the way forward amid diverse and often conflicting perspectives requires optimism and grit. Our primary motivation has been to demystify and reposition the act of design in the context of challenges that defy conventional paths of reasoning. Rather than claiming that design has all the answers, we assemble ideas and evidence about why and how design can be used to shape our emerging world.

We encourage readers to begin anywhere in the book, charting an individual path through narrative explications, informed opinions, and case studies. You are encouraged to synthesize your own thoughts and conclusions; to imagine, to question, and to allow new possibilities to emerge. To experience what for us has become the key lesson of our research: **to flow with, rather than fight against, complexity**.

Cambridge, Massachusetts
May 2025

"Moments of transformation aren't bad. We live for those moments. I mean, that's called hope."

Caroline Jones
Art historian and curator
Interview **32**

Frolic creates affordable, community-centered housing by partnering with homeowners to co-develop multi-family homes, fostering equity, shared wealth, and sustainable living.
MITdesignX 2018

CASE | **AFFORDABLE HOUSING**

Frolic Community
Building Equitable Communities

The housing crisis in the U.S. has many roots. Among them, land use is the most intractable. Across thousands of cities and suburban towns, post-World War 2 zoning and mortgage insurance programs encouraged the construction of single-family homes on lots one-quarter acre or larger — the American dream in action.

With much of their land built out at a low density, cities today are constrained from increasing their housing supply. Unmet demand has driven up housing costs, turning cities into enclaves for the wealthy. Add to this an aging population with a decreasing capacity to maintain and live comfortably in these big properties. According to AARP, around 80% of adults over 65 prefer to stay in their current homes and neighborhoods as they age (AARP, 2021). But for many, the only option is to move out to assisted living or subsidized housing — losing access to amenities, former friends, and familiar surroundings.

Founded in Seattle by developer Tamara Knox and environmental designer and planner Joshua Morrison and supported by a polymathic team of designers and financial specialists, Frolic Communities overcomes these obstacles using an innovative method of leveraging the value of a single-family home to underwrite development of additional affordable units on the same property. The homeowner may continue to live in their house or occupy one of the new units more suited to their needs. Each project is managed and designed as co-housing, in which neighbors may share common spaces and activities like cooking, childcare, and maintenance, thereby creating a truly mixed-income, intergenerational community. Implemented on an area-wide basis, the Frolic approach gently increases density over time while conserving the sense, scale, and social ties of the neighborhood. This is a powerful alternative to the displacement caused by gentrification or the wholesale redevelopment of the area for high-end housing. Frolic produces affordable units by minimizing land costs, as well as a unique finance-ownership structure enabling future residents to invest in the development of their own homes, reducing down payments to as little as $10K. The venture launched in Seattle and has rapidly expanded into other states, with a growing pipeline of projects working with municipalities and homeowners to expand housing opportunities without destroying neighborhoods, a model adaptable to any urban area.

2 Design in Time

Design is how we create change, and how we are changed by the world we create.

"Human beings have designed since the dawn of time. It's an innately human endeavor — people have always found ways to design things to make the world better, whether it's making an arrowhead or building a spacecraft."

Maria Yang
Mechanical engineer and product designer
Interview **67**

At its most fundamental level, design is a form of reasoning — one that works by synthesis rather than by analysis. It follows the logic of nature, which produces, responds, and adjusts, evolving increasingly advanced solutions to survival over time.

In collaboration with our colleague Francesco Zurlo, Dean of the School of Design of Politecnico di Milano, we have explored how the story of civilization can be read as a history of design. This becomes apparent when we investigate various "inflection points" throughout time — the moments when a particular discovery or paradigm shift radically alters the way reality is perceived and what is collectively understood as "truth."

In some instances, the design of a new object, tool, or system opens the way for new behaviors, relationships, and methods for applying the planet's resources.

Likewise, a philosophical shift, a meteorological event, or a change in food supply systems will adjust what is recognized as a problem and therefore what solutions are designed. In other words, each inflection point in human history turns upon how design affects the world and how the world affects design.

The "truth" of this interpretation is proved by its use. When we understand design as an innate human capability that enables us to synthesize an alternative future and thereby predict the next version of reality, we become intentional agents, empowered to shape the world we envision. Conversely, if we view design as merely an aesthetic endeavor to boost the appeal of objects or brands, we limit ourselves to a version of reality in which humans are reduced to passive consumers.

The failure to recognize the centrality of design to human life has resulted in conditions dangerously close to the second version of reality. However, in the current state of global crisis and supercomplexity, we cannot afford to be passive or fatalistic.

Design and *Technē*

The following snapshots in time give form to the relationship between design and the world as it is perceived and acted upon. As we see it, design enables the balanced interplay of humanistic and technical knowledge, first articulated in the ancient Greek concept of technē: the skillful application of an art to practical ends. From a neurological perspective, this can be understood as the synthesis of left and right hemisphere functions in the brain.

Since the 18th century, the technical view of reality has come to dominate the humanistic, disrupting the fine-tuned equilibrium of technē. When this imbalance becomes untenable, we see an eventual return to the designer's mindset in which "making" and "thinking" are given equal value as part of an interconnected whole. We are now experiencing just such a turn: It is time for the contemporary use of the word "technology" to return to its roots.

TOOLS

LOGIC

PERSPECTIVE

Design and Meaning

The points of inflection marked here are not intended to be comprehensive. Instead, they illuminate major shifts in human reasoning about the nature of reality and our place within it.

Design invests meaning into the human environment in ways that either reflect or reform those shifts in reasoning. For example, the orders of Greek and Roman architecture, tuned to human perspective and proportions, embody the values of

hierarchy and stability dominant during classical antiquity. To reapply these styles is to evoke those values — whether for progressive or retrogressive ends.

As illustrated by the following sequence, design will shape the course of human society, whether or not we take an active stance of conscious intervention. At our current crucial point of inflection, it is our choice which way we turn. There is still time to design.

REASON

MACHINES

MODERNISM

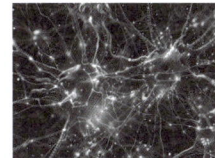

SUPERCOMPLEXITY

Tools

Blombos Cave, earliest known Homo sapiens drawing on tool; South Africa, ca. 100,000 to 70,000 BCE.

Homo naledi symbols in Rising Star Cave, South Africa (ca. 300,000 BCE).

Recent discoveries in archeology reveal that the capacity for design existed in proto-human species as early as 300,000 years ago in the use of shapes, visual symbols, or patterns to convey meaning (Berger et al., 2023). Patterns such as those engraved upon tools are remarkably similar across time and cultures, discovered in caves as far apart as South Africa, Gibraltar, and Australia.

Then as now, design may be described as the confluence of utility (the tool) and meaning (marks of the individual who made or used the tool). Such marks, or "drawings," provide a means to convey ideas and relationships that cannot be explained by the physical world. In design, the act of drawing (whether on rock, paper, or a digital screen) can be compared to the use of mathematics in science, or language when making an argument — ways of expressing and organizing what we know. Whereas mathematics is a closed system, seeking symmetry or components to equate, drawing is an open-ended instrument that thrives on intuition, asymmetry, and serendipitous emergences. Its impact on the world has been equally profound.

"Design is a cultural imperative, as essential to human survival as language, ritual, and myth."

Caroline Jones
Art historian and curator
Associate Dean MIT SA+P
Interview **32**

Logic

Aristotle

Over time, the advancement of language and numbering systems enabled aspects of the perceived world to be counted, categorized, and organized into hierarchies. These methods of ordering existence became a cultural paradigm during the centuries of ancient Greek civilization, culminating in the writings of Aristotle (384–322 BCE). Aristotle's dissection of reality is often considered a forerunner of the scientific method, given his insistence that facts must be founded upon empirical evidence — knowledge confirmed by the human senses and organized by logical reasoning. As framed by his "Four Causes," the reality we perceive can be understood by dividing the world into constituent parts, ordered by their intrinsic attributes and ultimate purposes.

The corresponding influence on design can be traced to Aristotle's "Ways of Knowing," in which a distinction is made between technē," knowing by making, and epistēmē, knowing by rational thinking (Parry, 2024). According to this hierarchy, "making" is considered less valuable than "theorizing" (Wang, 2013). The dichotomy laid the foundations for Western thought, which has repeatedly favored science as the primary arbiter of reality over the arts, humanities, and design. As we shall see, this paradigm is now being called into question, based on the findings of science itself.

Perspective

Between the 15th and 17th centuries, the rediscovery of classical Greek and Roman texts enabled a partial reconciliation of technē — resulting in an extraordinary era of cultural flourishing and a return to classically inspired architecture and design.

The rebirth signified by the word "renaissance" is illustrated in the reinvention of linear perspective by the Italian master metalsmith, engineer, and architect Filippo Brunelleschi. Standing in front of the Baptistery of San Giovanni in Florence in 1415, holding up a painting of the building in its surrounding context, Brunelleschi tested a mathematical system of representing depth on a two-dimensional surface. His technique — an advancement beyond the intuitive approximations used by the Greeks and Romans — passed the test (Lesso, 2022): The image matched the view precisely, demonstrating that a human tool — geometry — could reproduce what the eye sees.

Brunelleschi's achievement signaled the revival of technē as central to human invention. Writing 20 years later, Leon Baptista Alberti dedicated his landmark 1435 treatise De Pictura to Brunelleschi, crediting him with uniting art with science and mathematics — an act of synthesis arguably as decisive to the development of Western technology, science, and culture as the discoveries of Galileo or Copernicus (Edgerton, 2006).

Before and after Brunelleschi. Top: Medieval depiction of the Last Supper. Bottom: The Last Supper by Leonardo da Vinci (1495), in which Christ marks the vanishing point.

Before Brunelleschi, images on a two-dimensional surface presented foreground and background on the same plane; an ethereal and detached "God's eye view" of reality as opposed to the three-dimensional world we experience in the unfolding of everyday life (Kemp, 1990; Edgerton, 2010). Thereafter, representations of the world became more anthropomorphic, anchored to physical space as seen from the perspective of a human viewer; a world shaped by human perception (and human hands) as much as by the will of God and the divine right of kings.

Brunelleschi's seemingly modest achievement marked an overarching cultural transformation in Europe — encompassing not only art but an implicit repositioning of humanity's place in the world. The construction of perspective views became a powerful organizing principle in the design of cities (Kostof, 1991). When we look down the streets of contemporary Rome, Paris, or Washington DC, we see Brunelleschi's vision given physical form in the built environment.

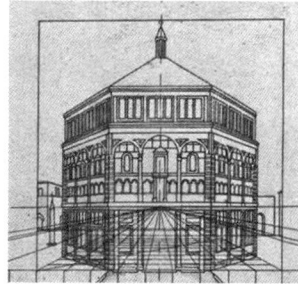

Diagram demonstrating Filippo Brunelleschi's perspective technique from a lost painting of the Battistero di San Giovanni. Kunsthistorisches Institut in Florenz, Max-Planck-Institut. ©2006, Scala, Florence / Art Resource, N.Y.

Rome today, Piazza del Popolo: Perspective in urban form. The obelisk, placed by pope Sixtus V in 1595 became the vanishing point in an urban perspective form that evolved over the next 300 years.

Reason

The empowerment of individuals and societies in relation to organized religion prompted the first breaks between church and state, and the advancement of scientific methods to decode the mysteries of the natural world. Truth was now defined by what individuals could observe and understand independently, forging a distinction between human beings and their environment. This philosophical shift laid the foundation for a worldview that grounded all aspects of life in observable, measurable facts.

The period between the 17th and 18th centuries, known in the West as the "Age of Reason," was ushered in by the philosophy of René Descartes, whose famous dictum "cogito ergo sum" has its origins in Aristotle's prioritization of human thought. Reached via a method of deductive reasoning, Descartes' statement might equally be read as "I think rationally, therefore I am." Implicit and intuitive ways of knowing were once again sidelined in favor of a quest for certainty and an emphasis on the instrumentality of cause and effect.

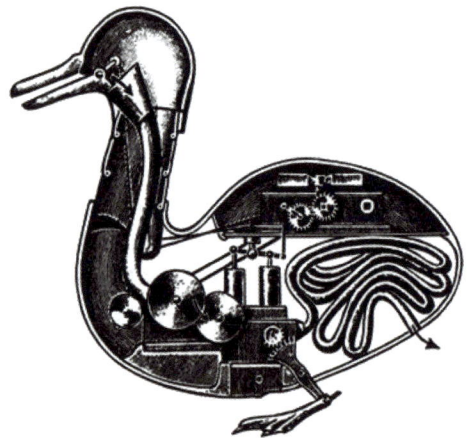

Vaucanson's Automatic Duck 1764.

Scientific empiricism was prioritized in the pursuit of truth, framing reality in terms of fixed facts that could be applied for specific ends. This was a world in which natural systems and human societies could be engineered; reduced to mechanisms that could be analyzed, understood, controlled, and even fixed by human intervention.

Meanwhile, the establishment of European colonies led to an influx of new commodities and aesthetic influences. This gave rise to a period of design characterized by embellishment and decoration across all forms of material culture, later counterbalanced by the return to neoclassical simplicity inspired by archeological investigation of the ancient Roman sites of Herculaneum and Pompeii in the mid-18th century. Reason, it seemed, had rediscovered its aesthetic counterpart in classical order — based on precedents ironically preserved by the chaos of volcanic eruptions.

"Design is a way of knowing that is not science and not the humanities. It is a different way of knowing, understanding, and creating the world. It allows us to balance, through a making process, intuitive conceptions of the world."

Elizabeth Christoforetti
Systems designer and technologist
Interview **17**

2—PHOTOGRAPH 2. EAST VIEW OF LOWELL IN 1838

Left: Early utopian view of Lowell, MA in 1838. Right: Literary publication by the mill girls — single daughters of New England farm families, attracted by the promise of secure housing, education, and payment in cash (sent back home).

Machines

The paradigm of mechanistic thinking inspired by Descartes paved the way for the Industrial Revolution in the 18th and 19th centuries; here, we see design playing a major role in the invention of new tools, manufacturing processes, and products that radically changed the organization and output of human labor and the reality of everyday life. The initial need to cluster machines in factories close to sources of water power led to a new form of city designed around the needs of industry, where thousands of workers toiled long hours to keep the machines running.

Over time, apparent progress gave way to exploitation in which workers were treated as inanimate parts of an industrial system geared for production and profit as economic disparity and poor working conditions were mirrored by environmental impacts. The shift from water to steam power dependent on fossil fuels led to the pollution of land, air, and water, as well as meteorological changes that marked the initial stages of global warming.

Mature Lowell. Boott Mill Yard,1884, with immigrant workers assembled before the bell; the mill was one of many mills stretching over a mile along the Merrimack River in Lowell.

Ten year old spinner in a North Carolina cotton mill, captured by Lewis W. Hine (1900 - 1937).

Daniel Burham's City Beautiful Plan for Chicago, 1909, inspired similar designs in cities across the U.S., hoping to bring culture, cleanliness, and health to utilitarian, industrial cities. Note City Hall, obelisk and radial perspective, neoclassical architecture, emphasizing order, beauty, and social cohesion.

Modernism

Early industrial cities, such as Lowell, Massachusetts (1822) and New Lanark in Scotland (1800), sought to prioritize worker welfare as part of large-scale industrial production (Crawford, 1995). However, when economic demands surpassed humanitarian values, the industrial city became associated with social decline and environmental degradation.

Calls for reforms in labor, living conditions, and public health during the latter half of the 19th century underpinned the emergence of the City Beautiful Movement in the U.S. and Garden Cities in the U.K. — manifestations of a new discipline of city planning that aimed to promote human welfare and social cohesion by designing cities that were healthy, well ordered, and aesthetically pleasing. Likewise, the new discipline of industrial design centered on producing mechanized

"Designers have moved in a very narrow scope of the practice... One of the most important things we can do is expand the vision of what can be done by designers."

Sofía Bosch Gómez
Designer and researcher,
social change
Interview **13**

In Europe following World War 1, modernist reformers rejected past cultural references seeking to establish a new social order epitomized by standardized, stripped-down products, buildings, and cities, made affordable to the masses. Proponents such as the Italian Futurists, and the German Bauhaus reimaged virtually all of the human habitat as one in which "form follows function." Below: The Bauhaus in Dessau and one of its most famous products, Marcel Breuer's Wassily Chair.

consumer products — from cars to vacuum cleaners — that were affordable, user-friendly, and appealing.

In the early decades of the 20th century, these concepts informed modernist visions of the future city produced by architects and industrial designers in the United States; in Europe, the Bauhaus movement aimed to simplify all elements of the material world to facilitate mass production and affordability. European and American ideals coalesced in the 1939 New York World's Fair: "The World of Tomorrow." The most popular exhibit at the fair was General Motors' Futurama, where 25 million people experienced a streamlined prediction of 1960 — a vision of nation-spanning interstate expressways, electrified homes, high-rise living, and expansive parks that replaced "outdated business sections and undesirable slum areas" (Bel Geddes, 1939).

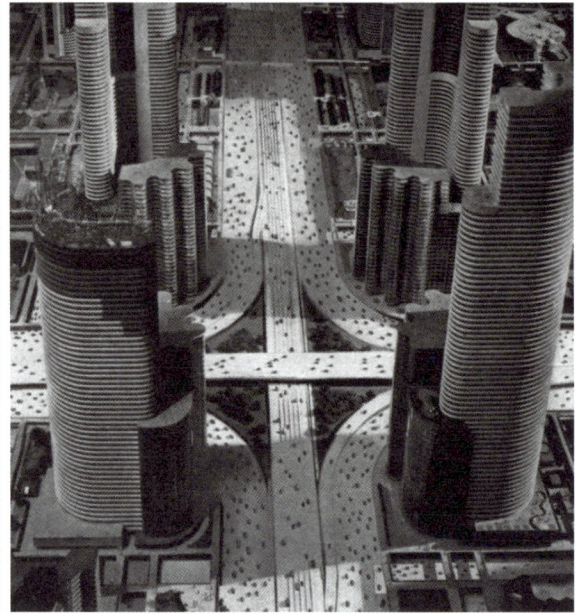

At the 1939 New York World's Fair GM Futurama, industrial designer Norman Bel Geddes fused modernist thinking in Europe and the U.S. to produce a spectacular vision of the future in 1960 with rebuilt cities linked by a national system of superhighways. Visited by 25 million people, the exhibit introduced narration synced to movement over a vast 3-D model simulating an airplane trip across the U.S. So powerful was the vision, that it led to federal housing, urban renewal, and interstate highway programs following World War 2.

Such visions had an extraordinary impact in the years following the Second World War, becoming a model for renewing devastated cities and progressive urbanism internationally (Johnson & Hitchcock, 1932) — a legacy that continues today. Levels of poverty, disease, and income inequality were significantly decreased as a result of the modernization of underdeveloped regions. However, underlying issues of pollution, resource depletion, and disparities in health and income persisted, accelerated by extreme urbanization and the centrality of the car as a preferred means of transport. Then as today, local culture and environmental consequences were overlooked in favor of universalizing ideals of progress.

"Design moves back and forth between intuitive experimentation and systematic problem-solving — between asking 'What if?' and 'Will this actually work?'"

Nicholas de Monchaux
Designer, author, educator
Interview **24**

Supercomplexity

Over the course of the 20th century, this vast transformation of the human habitat prompted an ongoing debate that set the technical dimensions of design against the humanistic. In design schools and professional practice, an analytical approach was once again prioritized over the synthesis of *technē*, even leading to attempts to reframe design as a science.

However, in recent decades, the flaws and oversights of modernism have risen to the surface. The condition of supercomplexity, outlined in Chapter 1, has meant that it is no longer possible to rest upon former ideals. Indeed, we are witnessing the disruption of modern thought and society at all levels. Contemporary technology — often divorced from the humanistic qualities that enrich the concept of technē — is evolving faster than social and legal practices, even before we can judge the impact on human relationships and our environment. The stakes are greater every day.

The point of inflection we term "supercomplexity" is marked by what has been referred to as the VUCA paradigm: volatility, uncertainty, complexity, and ambiguity (Bennet & Lemoine, 2014). This framing of reality provides instructive parameters for design, inviting an approach to problem-solving that requires a critical understanding of contexts, an openness to envisioning alternative futures, and a response to rapidly changing circumstances.

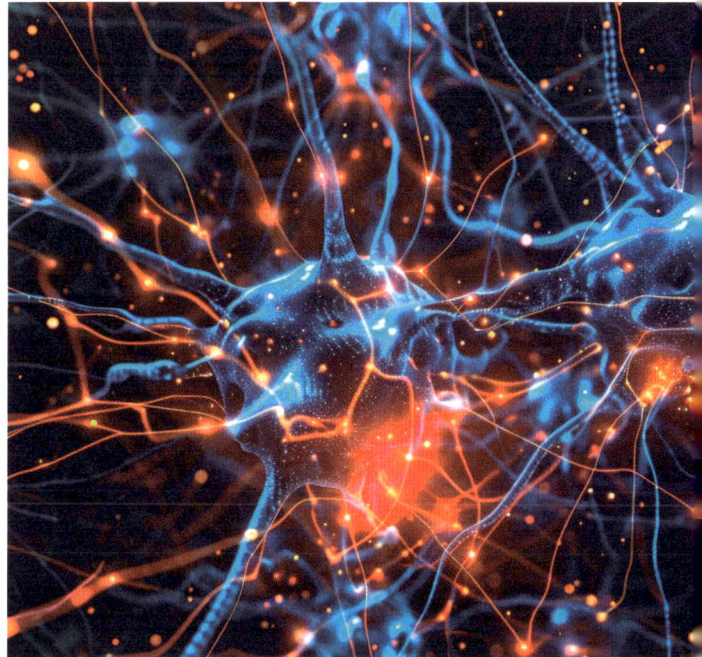

Future inflections

"Design with
tomorrow in mind,
not only today's
demands."

Roberto Verganti
Scholar and author
Art and innovation,
design theory
Interview **63**

The next inflection is already being set in motion
by design, a process we will explore in depth in the
following chapters. Arguably, this is where we will find the
future. Design is distinguished from other disciplines in
its unique approach to positioning a vision — rather than
a pre-existing fact — as its primary point of orientation.
In this way, design serves a predictive function, and
predictions affect actions in the present, whether or not
the vision comes to pass.

The predictive visions of the late 19th century — the
science fiction of Edward Bellamy (1888/2009) or
Jules Verne (1863/1996), for instance — and the many
manifestos of early 20th-century modernism — the
Futurist Manifesto, the Bauhaus concept of universal
design, Le Corbusier's *The Radiant City* (1933/1967) —
profoundly reframed perceptions of what was possible.
They have provided the elements, if not the exact
blueprint, for the innovations and social shifts that have
subsequently emerged.

Design is not an oracle, but it is a gauge by which the
future can be read. Look to what is being envisioned
and what actions are being taken to achieve that vision,
and you will gain a sense of where the world is turning.
That process of envisioning and revealing a pathway
between one world and another is explored in the
chapters that follow. The future is neither fixed nor wholly
indeterminate. It is the measure of our minds and the
capacity of our hands.

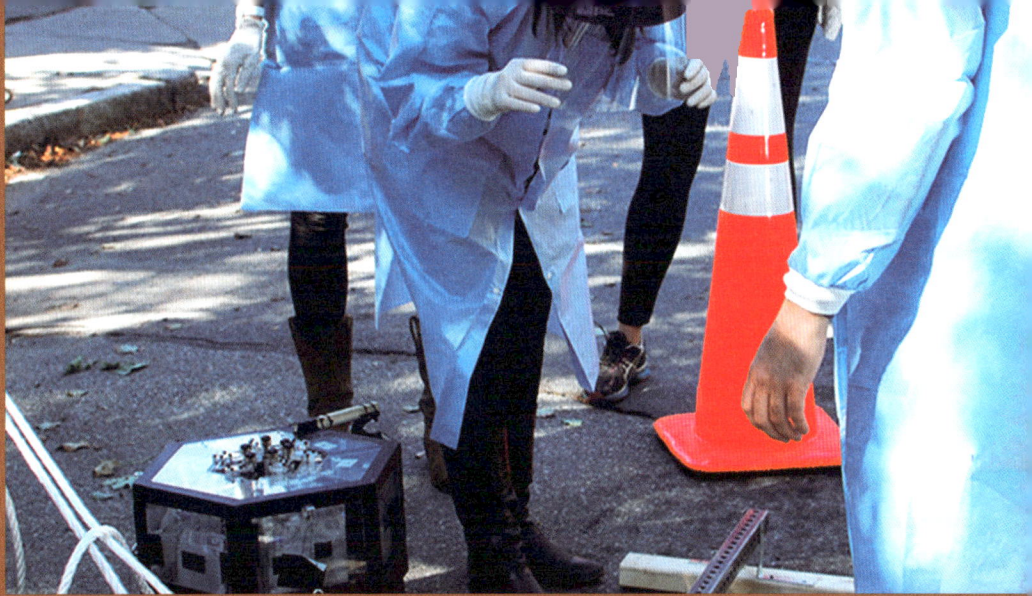

Biobot transforms wastewater into actionable data to protect and improve public health. MITdesignX 2017

CASE | **PUBLIC HEALTH**

Biobot
Improving Global Health

Since 1854, when John Snow tracked the source of cholera to a contaminated well in West London (Snow, 1855), epidemiologists have struggled to understand the scale and spread of disease in cities. The supercomplex interactions of people, surfaces, water, air, vermin, and waste thwart traditional indirect methods, such as contact tracing, spot testing, or counting hospital admissions. As cities increase in scale and density, the problem is multiplied (Hassell et al., 2017). Envisioning a new way to monitor public health, biologist Mariana Matus and architect Newsha Ghaeli founded Biobot Analytics to develop and deploy sampling and analysis technology for human waste in sewage. Using molecular biology and data science, Biobot transforms wastewater into population-level health data, where their sampling system functions as tiny "public health observatories," generating maps and reports that track the presence and trends of pathogens across different sectors of a city.

The company's first product was to measure and map drug metabolites in sewage to estimate opioid consumption. The data enables local governments to assess the scope of the opioid epidemic, allocate resources, and gauge the effectiveness of responses, thereby empowering communities to tackle the spread of disease in real time. Biobot rose to prominence during the COVID-19 pandemic, demonstrating that conventional methods were underestimating the spread and intensity of the disease.

The company has served over 700 communities, representing over 100 million people, and is reporting on multiple health threats — from diseases such as the respiratory syncytial virus (RSV) to foodborne illnesses and the presence of high-risk substances. Highly accurate yet relatively inexpensive, the system has the potential to be deployed anywhere as an integral part of a city's wastewater infrastructure, enabling health conditions to be monitored and diseases to be traced in real time, locally and worldwide.

61

3 Progettare

Progettare is the art of shaping visions into tangible realities.

"'Cultura del Progetto' means understanding design not as aesthetics, but as a sociological, historical, and cultural superstructure. It's about designing with awareness of context."

Stefano Mirti
Architect, educator
strategic design
Interview **42**

"'Progettare is a deeply human response to complexity that demands imagination and collective commitment."

Gianandrea Giacoma
Psychologist, psychotherapist, and consultant

How can we meet the demands of unexpected challenges and develop solutions better suited to a world of accelerating complexities? *Designing the X* proposes that we already possess the tools to effectively engage supercomplexity and proactively shape the future.

In the search for solutions in supercomplex environments, it is important to resist the urge to systematize and oversimplify. Such reductionist approaches seek to minimize complexity by identifying the key elements of a problem that can be related or modeled in a mathematical equation. During the process of simplification, some aspects are inevitably left out; these elements may turn out to be more important than what has been included.

The design concept of *progettare* offers an alternative approach. In Italian, the word can be understood as the art of shaping visions into tangible realities. However, progettare is not as simple as a "plan" that involves following a preconceived set of steps toward a singular solution. As our colleague, psychologist Gianandrea Giacoma, observes, "there is a natural tension within the word that relates to the motion of 'throwing forward, to project an idea, an intuition, a vision."

Progettare involves recognizing that the movement toward a desired future outcome is constantly shaping and redefining what the outcome might be.

This is an accepted phenomenon when making art. A drawing is simply "a line going for a walk," as Paul Klee noted in his *Pedagogical Sketchbook*, (Klee, 1953); for Michelangelo, the process of creating a sculpture involved gradually uncovering a form hidden within the marble (Kárpáti, 2019).

A Walk with the Child; Paul Klee. In the digital collection of the University of Michigan Museum of Art.

The subtle relationship between vision and realization has likewise been articulated by Dennis Gabor, Nobel Laureate in Physics, whose invention of holography demonstrates his own instinctively spatial understanding of visualization. Futures are found not only through observations of trends and scenarios but also as an outcome of discovery and imagination:

The future cannot be predicted, but futures can be invented. It was man's ability to invent which has made human society what it is. The mental processes of invention are still mysterious. They are rational, but not logical, that is to say not deductive. The first step of the technological or social inventor is to visualize, by an act of imagination, a thing or a state of things which does not yet exist, and which to him appears in some way desirable. He can then start rationally arguing backwards from the invention, and forward from the means at his disposal, until a way is found from one to the other (Gabor, *Inventing the Future*, 1964, p. 161).

By taking on the mentality of design, innovators gain a key advantage. Designers are schooled in the practice of navigating uncertainty — the void that humans instinctively shrink from, the gaps in knowledge that we crave to close.

The impulse to close the gap propels us toward discovery and invention, but it can also lead to restricted thinking. Designers apply a different mindset, embracing uncertainty as an essential ingredient for effective problem-solving — guided not by preconceptions, but by the evolutionary nature of the design process itself.

Engagement Principles

The difficulty of moving forward amid uncertainty should not be underestimated. It takes nerve to reject the comfort of a fixed endpoint (a single correct solution) and the security of a well-trodden path from A to B. Designers learn to stay on track by respecting key strategic principles or "guideposts."

The interviews conducted for this book revealed several overarching principles to engage supercomplexity, drawn from the experience of innovators applying the lens of design across multiple areas of practice (see Figure 3.1).

These engagement principles underpin the themes discussed below and the journey through the design process outlined in Chapter 4:

1. **Contextual** — supercomplex problems materialize within the unique local circumstances of their place and time. They can neither be generalized from other situations nor understood and addressed by applying abstract laws.

2. **Polymathic** — such problems transcend the domain of any single discipline. Working with supercomplexity demands knowledge of diverse fields, necessitating the involvement of specialists and stakeholders with multiple perspectives and expertise.

3. **Evolutionary** — solutions cannot be reached via a prescribed set of steps. Rather, they evolve, emerging from a spectrum of options to be tested against the vision of a future that is conceivable — if not yet feasible — in the present.

The three principles are interdependent. Understanding the holistic context and its ongoing development requires polymathic thinking, leading to the evolution of approaches that serve to better fit, or actively change, the context. Applying these principles in a nonlinear way at multiple scales can guide problem-solvers in navigating the unknown and unexpected, leading to new insights, emergences, and long-term solutions with their own inherent dynamism.

INTERVIEW FINDINGS

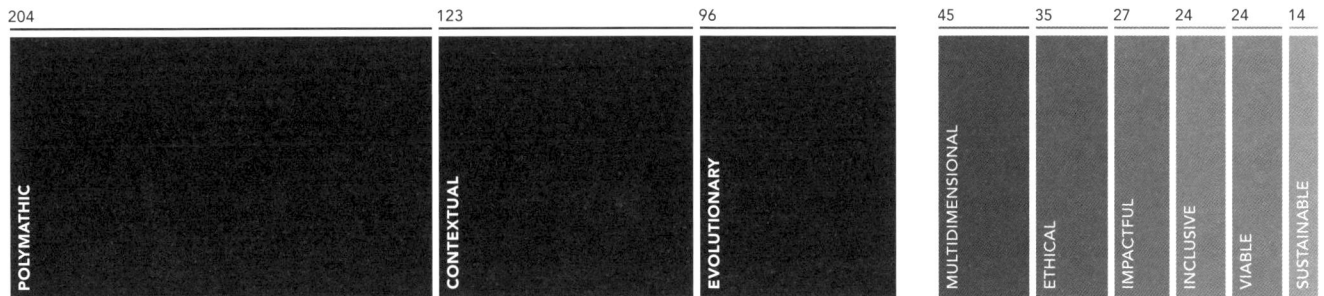

Figure 3.1 | Engagement Principles. Data analysis of interview transcripts revealed Contextual, Polymathic, and Evolutionary as key principles to engage supercomplexity. Numbers in the graph reflect ranking of concepts based on their frequency and contextual significance across interviews for this book (see Appendix, Methods).

PRINCIPLE 1 | CONTEXTUAL

The contextual understanding of a problem requires taking account of the time, physical place, culture, environment, and social dynamics that influence potential solutions. It acknowledges that beliefs, behaviors, and outcomes vary depending on where and when they are situated, forming conditions that constrain or promote what is possible.

By acknowledging human thoughts, desires, and emotions as equivalent to measurable facts, design is both the inverse and the complement of reductive analysis, which deals with the facts alone. This acknowledgment has profound implications for engaging supercomplexity.

Against Abstraction

The impact of supercomplex problems manifests locally, while these problems are also deeply interconnected with global systems. This is what makes supercomplex problems so confounding; unless a solution aligns with the local physical and social environment, as well as broader global forces, it is unlikely to succeed.

Generalized, abstract models and solutions derived through analysis (which may be valuable to disembodied technical applications) often fall short in the context of a specific place and time. A more flexible and adaptable approach is required.

"The evaluation of good design is not absolute. In evaluating, we must relate contexts, conditions, and outcomes."

Alberto Bassi
Design historian and critic
Interview **7**

Central Artery, Boston. Built in the 1950s, The U.S. Interstate Highway System met abstract federal engineering standards in rural areas but caused significant issues in cities, where due to cost and physical constraints (or discrimination), highways ended up being routed through poor, ethnic neighborhoods and declining industrial waterfronts. This led to environmental and social injustices, which the recent U.S. Infrastructure Investment Act aims to redress by removing or redesigning these routes to restore communities and access to invaluable natural assets

"Engaging with complexity means immersing yourself in context — observing, listening, identifying unmet needs — sometimes articulated, sometimes unarticulated — looking for hidden opportunities for innovation, then defining where meaningful solutions can emerge."

Peter Coughlan
Strategic designer and systems innovator
Interview **20**

Design synthesis is well suited to this need, as it seeks to invent ways forward within a local context — now and in the future. In contrast, analysis works backward from present conditions to interpret what is occurring at the most fundamental, abstract level. The conclusions drawn from abstract thinking may be far off the mark when applied to a specific environment, culture, and time; local conditions can produce radical divergences from the path anticipated by the abstraction.

Beyond User-Centric Thinking

Designers learn to explore and understand the context not only from their own vantage point, but also from the perspectives of other disciplines and stakeholders that the project, service, or policy is intended to serve. However, it can be reductive to assume that an "ideal user" is central to the solution. This simplification is

Figure-ground diagram: a vase or two faces?

appropriate if the task is to design an object like a shed or a spoon, where cause and effect are evident. However, moving into supercomplex problems — such as mitigating global warming, managing human migration, combating pandemics, and designing an autonomous mobility system — requires designers to fit the system into its social, technical, and physical circumstances.

In such situations, finding an ideal user — the reductionist surrogate for a human response — may be difficult or impossible. It is more likely that there will be multiple user types with different stakes in the outcome.

The aim must be to evolve both human and contextual systems simultaneously toward a more productive fit. This is modeled in the natural world, where life has shaped, and has been shaped, by planetary conditions on Earth; from the emergence of oxygen-emitting bacteria 2.9 billion years ago (Fournier et al., 2021) to the current human-induced climate crisis.

The physical environment is correspondent with living systems, like a figure-ground diagram in which objects and background are of equal importance when grasping the whole. In fact, by simply focusing on one or the other, we may perceive entirely different wholes.

Local and Multiscale

Context cannot be generalized, but it can be influenced. To achieve this, designers start by envisioning the potential impacts of an action at a local level, taking into account the conditions that shape understandings or applications in the given circumstances. This demands the deep involvement of stakeholders, as they are an integral part of the context.

"I personally find user-centered design unethical. It is essentially about consumerism. My hope is to shift the conversation from mapping individual desires to a more value-sensitive approach where stakeholders are the primary drivers behind design research. We must ethically relate to their context."

Elizabeth Christoforetti
Systems designer and technologist
Interview **17**

While all supercomplex problems are indeed local, in some cases their context may extend to be regional, global, or even universal. In the policy arena, there is a tendency to group such problems together, for instance as typical of urban or "third world" issues. Such characterizations encourage generalized disciplinary approaches and abstract models that are not well fitted to any specific place or time.

Thinking Contextually in Supercomplexity

As the world increases in complexity, so does the rate of change. While a context may seem static from a human time scale, looking at "then and now" pictures of the same location in any city is enough to dispel the perception that context is fixed. Such visual comparisons emphasize the built figure of the city, but there are equally profound changes in the ground: different climates, economic systems, patterns of movement, and judgments of value.

Changing contexts leads to different perspectives on problems and opportunities. What is considered impossible at one point in time can be commonplace in the next. Hence, options for solving supercomplex problems must constantly be tested for their fitness, not only to the present but also to the envisioned future.

Left: Upper Broadway Boulevard, New York City, ca. 1885. Right: Upper Broadway today, 2025 (Google Earth).

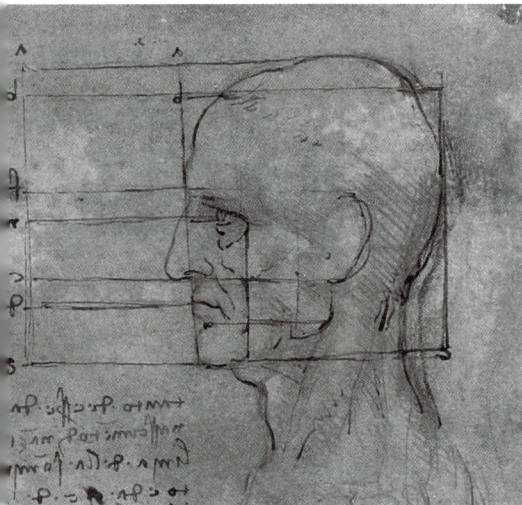

Mathematical proportions of the human head: "Technē" (Leonardo da Vinci, 1490).

"Design isn't only a solution-maker, but also a question-raiser."

Kengo Kuma
Architect
Interview **36**

PRINCIPLE 2 | POLYMATHIC

A paradigm of divided disciplines has shaped culture in the West for millennia. Over time, the many disciplines recognized today — from architecture to zoology — have evolved their own problem sets, theories, methodologies, and specialized language, enforced by rules or codes of professional behavior. Arguably, the origins of this framework can be traced back to the 4th century BCE, when Aristotle classified the "Four Causes" — each a way of answering the question "why?" (Reece, 2018).

The rise of analytic reasoning is simultaneous with the segmentation of knowledge into distinct categories, resulting in standardized rules of investigation and practice in the world. The word "discipline" has thereby taken on a dual meaning, indicating both a field of knowledge and a way of adhering to the rules.

This siloed approach has served to deepen knowledge within the narrow frame of each discipline. However, when the heuristics (conventional problem-solving techniques) espoused by each discipline are codified as laws, whether articulated or unconsciously accepted, these ways of effecting change in the world become very difficult to challenge or modify.

As the challenges we encounter become increasingly interdependent, the static segregation of disciplines serves to limit perception and foreclose understanding of a specific context and alternative routes to a solution.

"Design transcends scale, disciplines, and domains. We must combine science, technology, and human agency to co-create new hybrid fields, cultivating polymath designers who build what's next."

Skylar Tibbits
Designer, computer scientist
Interview **58**

We propose an alternative "polymathic" approach: one that weaves together multiple branches of knowledge.

Individuals referred to as polymaths are considered fluent in many fields and professions. They follow historical precedents set by the likes of the Renaissance painter, scientist, inventor, and architect Leonardo da Vinci, or W. E. B. Du Bois, whose legacy as a writer, sociologist, historian, economist, and activist continues to resonate in the civil rights struggles of today. Aristotle, whose inquiries ranged from the natural sciences to the arts, was likewise one of the foremost polymaths.

We attribute the term "polymathic," as applied within the practice of design, to our colleague Skylar Tibbits, founder and co-director of the Self-Assembly Lab at MIT. To be polymathic is a mindset as much as a practice. It describes an outlook of synthesis that motivates designers to move beyond the confines of disciplinary knowledge, seeking instead to enhance an understanding of the interconnected and dynamic processes that characterize life on this planet.

We would argue that design, as a process of synthesis, is inherently polymathic. To gain a comprehensive understanding of the context of the problem and to fulfill the requirements of the design team's vision, it is necessary to mediate between different fields, prospecting for diverse knowledge and seeking input from multiple areas of expertise.

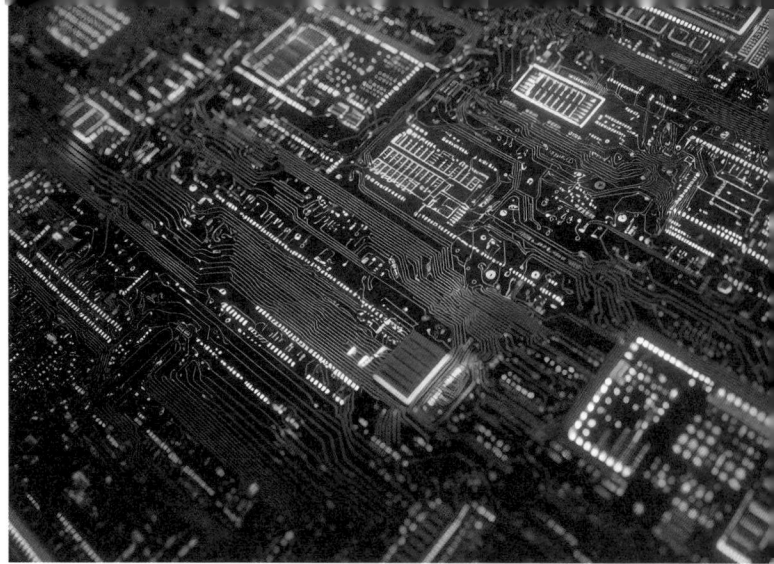

This results in the built-in recognition that design is not a standalone discipline but one part of a larger whole.

The polymathic principle also reveals the necessity and pragmatism of recognizing that stakeholders should have equal footing as the various disciplinary experts and designers. Stakeholders are experts in the lived experience of a place; to fail to recognize their first-hand understanding would dramatically limit the resources available for addressing the problem.

By taking this approach, we advocate for designing *with* stakeholders in a process where solutions can be proposed *by* stakeholders; a significant advancement from broad (and arguably objectifying) campaigns to solve problems *for* abstract "users." Designers must be adept at cultivating trust and building a sense of personal investment among those positioned to benefit from a proposed solution, a process we explore in more depth in Chapter 4.

Engaging Supercomplexity

The ongoing emergence of supercomplexity further highlights the need to move beyond singular disciplines. Contemporary challenges involve a multitude of diverse human and non-human systems, stakeholders, agendas, and contextual variables that are difficult or even impossible to model.

Consider a city, a sports team, or an international corporation where intangible factors such as cultural values, emotions, and aspirations can affect the performance of the whole. For instance, in evaluating urban performance, it is often considered sufficient to quantify and analyze measurable factors such as building conditions, traffic volume, and economic value. However, it is increasingly clear that this approach does not encompass the full scope of interconnected factors; what ranks as a high-performing city in the data may actually offer a poor quality of life.

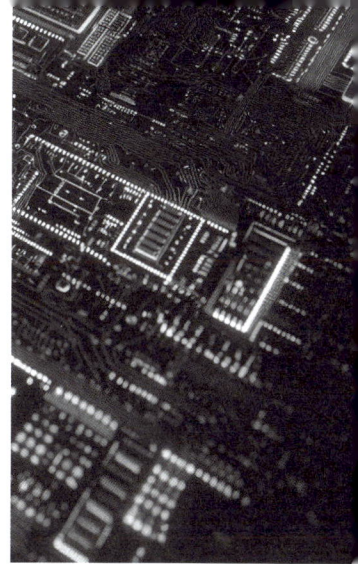

The ever-evolving nature of supercomplexity further emphasizes the need to encompass multiple perspectives on a problem. Changes in one system can have cascading effects on others, leading to unfathomable feedback loops and behaviors. The engagement of one aspect of a problem has a ripple effect on the whole, resulting in overlapping layers of evolving complexity. This recognition has led to the definition of new cross-cutting areas of study and practice: biochemical engineering, health informatics, and behavioral economics, to name just a few. Indeed, as knowledge accelerates exponentially, we are witnessing an explosion of hyphenated fields to the point where the word "discipline" is losing its meaning.

Human society and the media have yet to catch up. Today, socioeconomic pressures and systems of higher education tend to pigeonhole emerging professionals into a single career path, even as educators promote the benefits of being a "well-rounded" individual. While this characteristic might be appreciated in terms of psychological development and the ability to form strong relationships, the primary arbiter of academic success is less expansive. The subliminal mantra (and trap) is to "do one thing and do it well."

"In its best version, a designer looks at the facets of human experience. And in solving a problem, looks for a way to address and weave together as many of those as possible. It doesn't matter if you're designing a pencil or laying out a city."

Jeff Klug
Architect and educator
Interview **34**

75

Enriching the Picture, Expanding the Frame

It's no surprise that supercomplex real-world issues tend to emerge at the intersection of disciplinary boundaries, indicating the increasing obsolescence of conventional knowledge structures. While these structures previously provided a pragmatic framework for organizing understanding and action in the world, it is vital to be aware of when a frame is no longer beneficial.

During our research for this book, we have been encouraged to see the signs of an emerging shift in consciousness. With new knowledge gained from a spectrum of disciplines ranging from science to sociology, many observers are recognizing that the old boundaries of disciplinarity are giving way to a more holistic view of nature and the human context. The human picture of the world is growing richer, and the frame we develop must be just as expansive.

Polymathy helps to achieve this objective. As a method and a mindset, it is not merely an accumulation of facts across various fields, but a deeply interconnected understanding that simultaneously sees the bigger picture and the finer details. It is a way of shaping a process that opens pathways to effective and sustainable solutions. These are achieved by:

Highlighting related patterns of thinking across time, disciplines, and scales, in ways that reveal underlying connections that can be applied to solve problems and drive innovation.

Encouraging a team process of engagement that incorporates multiple perspectives, priorities, and criteria of success.

Recognizing stakeholders as experts in the context and systems of a specific place and time or the local expression of a supercomplex problem.

Survive and Thrive

It's no overstatement to claim that polymathy has become a critical survival skill. In a world increasingly defined by supercomplexity and an accelerating pace of change, conventional practices fall short. However, a polymath will never draw a blank. By navigating the unknown and unexpected, they can tap a vast reservoir of insights and skills across multiple fields.

This affords polymaths the agility to pivot between different domains, identifying strategies that evolve in response to new information and changing circumstances. Proposed solutions have the potential to remain relevant and effective, even as the external environment is in flux. It is comparable to having a key that fits many locks, disclosing potential outcomes that are hidden from narrowly focused approaches.

Additionally, a polymathic mindset encourages continuous learning and curiosity, traits that are vital for staying ahead and remaining proactive rather than reactive. Indeed, by allowing for greater complexity in an organizing framework of the world, a polymathic mindset ultimately allows for greater

"When designing for extreme environments, everyone needs to be at the table."

Dava Newman
Aerospace biomedical engineer
Interview **46**

order and resolution. It can reveal patterns where others see only chaos, anticipating challenges before they become apparent, and enabling the crafting of holistic, integrated, and long-lasting solutions. Complexity is no longer perceived as a barrier but as fertile ground for inventive outcomes.

Resistance and Receptivity

To be a polymath is typically considered a positive trait. However, in practice, it has prompted resistance from traditional disciplines. Scientists and designers have been among the first to resist — all the more surprising, given that these practices are arguably the most inherently polymathic. The more detailed and granular the data identified through scientific investigation, the more specialized the field has become. Designers, meanwhile, have been blinkered by popular mythologies of the singular visionary thinker who conjures a solution through sheer creativity.

However, challenges to traditional disciplinary thinking are emerging from within each field. For example, it is increasingly accepted that scientific facts cannot be established free of human prejudice and perception (May, 2021). Design, conversely, is not reliant solely on creative instinct, but mediates between different disciplines to generate and refine emergent ideas. Both fields, alongside many others, are recognizing the benefits of working toward a vision of the wider whole, looking to polymathic practice as a way to deepen understanding and effectively engage supercomplexity.

PRINCIPLE 3 | EVOLUTIONARY

An evolutionary process can be understood as the unscripted development, change, or transformation of a system — whether an organism, a technology, or a concept — through a process of selecting among multiple options over time.

As an approach to problem-solving, the evolutionary principle highlights the need for anticipating change, even radical change, as part of a trajectory of development. It also recognizes that solutions and contexts co-evolve. The process is one of continuous and non-linear decision-making in support of the vision, which may also adapt as new insights and challenges arise. In this way, the forking paths of each choice shape the way to the preferred solution.

Biology and Beyond

In the discourse of biology, evolution refers to changes in the heritable traits of living species over successive generations (National Academy of Sciences). This occurs through a process of random genetic mutation in which "optional" traits emerge. The options best suited to the context are naturally selected to survive, while others die out (Caporale, 2003). Over time, this process can result in profound changes to a biological system, including the emergence of unforeseen elements and relationships that cause one system to succeed and even dominate over others. As a process of synthesis, evolution does not follow a prescribed model. Instead, it exploits opportunities as they are presented.

Such processes in nature are often referred to as "survival of the fittest" (Darwin, 1869). However, as many theorists have argued, the concept can also be applied to the evolution of the physical universe. In the *Emergence of Everything* (Oxford, 2002, p. v), biologist and philosopher Harold J. Morowitz describes 28 emergences that have resulted in the conditions of life on Earth today: from the explosion of unformed energy at the instant of the Big Bang, to the emergence of elementary particles, atoms, molecules, stars, planets, living organisms, language, and human consciousness (Morowitz, 2002). Morowitz's research also demonstrates that evolutionary processes govern domains beyond biology, shaping the development of technologies, societies, and systems of all kinds. In each of these cases, evolution as emergence depends on the presence of diverse options and the specific nature of the selector — whether gravitational attraction, survival, or (in the context of design) an envisioned future state.

Adaptation and Evolution

The evolutionary principle can likewise be applied to the development and transformation of customs, beliefs, institutions, or behaviors in response to new challenges and shifting values, incorporating new knowledge and techniques over generations. For example, the vision of universal suffrage and representative democracy that emerged during the 17th and 18th centuries continues to evolve today.

At root, it is the nature of dynamic systems to undergo modifications or improvements over time in response to changing external environmental conditions or internal pressures. Designers must therefore take the time to discover, even invent, different perspectives and optional ways forward. This necessitates an openness to emergences as they arise and an awareness of the fact that there may not be a singular optimal answer.

Unlike mathematics, in which missing elements are identified to balance an equation, design has no such ledger. Faced with changeable conditions of context and stakeholders, it can be difficult to know when we have solved a problem. This can be frustrating to those schooled in analysis, where the objective is to find the "right" answer. The designer recognizes that several solutions may be equally effective, depending on the vision they seek to achieve.

Instead of attempting to solve a problem at the outset, design evolves toward an envisioned state over multiple cycles, incorporating feedback, addressing limitations, and adapting to new information. Constraints are essential to focusing the design process. They are the methods we use to narrow and accelerate the flow of synthesis toward a desirable outcome, a process discussed further in Chapter 4.

"Most business decision-making is stationary or static until proven otherwise — whereas design assumes change and evolution as the norm."

Melissa Marsh
Workplace strategist, change management
Interview **41**

Vision as Natural Selector

Taking an evolutionary approach involves decision-making among options, a process guided by keeping the larger vision in focus as each small choice is made. This course of action involves testing and retesting multiple options at different scales against an imagined future, incrementally evolving a path forward. The eventual preferred solution may not be perfect; however, in its context, time, and place, it can emerge as the pathway most likely to align with the vision.

In design, the vision serves as a "natural selector" at all scales. It expands the notion of context to include not only what is measurably real (verified by hard data), but also what is imagined — the vision of a desired future state. By granting these two types of knowledge equal value, a richer array of potential solutions emerges. The only way to predict the next emergence is to envision it; the future cannot be modeled using data from the existing situation alone.

Ultimately, emergence occurs when the product begins to exceed the sum of its parts, recognized by a leap toward the vision that could not have been foreseen. The process of evolution challenges our expectations of what is possible, even leading to outcomes that might seem bizarre or laughable — like a fish with legs.

"The design process makes the appropriate connections between all things. It is a natural selection process."

Taku Satoh
Graphic designer
Interview **55**

Tiktaalik lived 375 million years ago. Its feet, which evolved from lower fins, helped it to sense food in the sediment beneath the water, with the added benefit of enabling it to scout the shoreline (University of Chicago, 2014).

Uncertainty and Reality

The richness of non-linear creation is manifested by the extraordinary diversity of life on this planet, from human beings to the smallest of microbes — all interconnected within the Earth's biome.

The true power of design arises when working with natural processes of synthesis for the purpose of effecting practical and meaningful change. As we have stated, this is not a capacity afforded to just a few empowered individuals. It is an integral function of the human mind; as we perceive, plan, and make, we are all designers.

This is inferred by neurological studies showing that every human action or perception is first envisioned (and therefore, "predicted") in the right hemisphere of the brain before being operationalized by the left. The perception of reality — what is perceived to be true or possible — is likewise shaped in this process (Clark, 2024).

The brain is continuously assessing and predicting thoughts and actions based on prior experience. In the process of perceiving the world, we are wired to construct narratives that align with our own expectations, applying sense data to confirm or adjust the prediction and potential viability of a thought or action.

The tendency of the brain to alter perception by filling in gaps based on past experiences or a projected desire means that the reality we perceive is partly of our own design. Apparent "reality" is not fixed but corresponds to the way we assess and adjust our attention and perception. According to neuroscientist Patrick Cavanagh, research professor at Dartmouth College, "We're seeing a story that's being created for us. Most of the time, the story our brains generate matches the real, physical world — but not always. Our brains also unconsciously bend our perception of reality to meet our desires or expectations… [Such] illusions present clear and interesting challenges for how we live: How do we know what's real?" (Resnick, 2020).

Such phenomena challenge us to continually reassess our framing of the world with greater humility, reflection, and responsibility. The growing understanding that reality is individually perceived and collaboratively shaped has significant implications for engaging supercomplex problems through design:

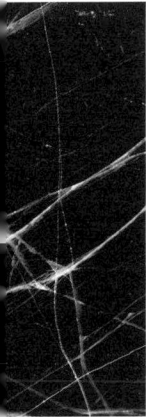

First, by giving credence to the act of envisioning, which can be seen as a refined application of an instinctive human function

Second, by calling into question the validity of accepted "facts" that shape individual and collective perception of problems and potential paths to solution

Third, by reassessing the primacy of scientific analysis as the only reliable basis for knowledge, looking instead to methods informed by multiple modes of perception and ways of knowing.

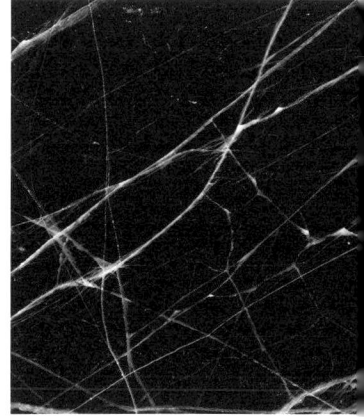

This realization places design synthesis and scientific analysis on an equal footing in their potential to affect the world: They are complementary practices of progettare. As we collectively "expand the frame," questioning perception and opening new pathways for innovation, the "bigger picture" is revealed to be richer and more nuanced. This vision emerges through processes of synthesis, with design serving as a key to unlock the potential of each area of expertise.

Designing for the Unknown

As discussed throughout this chapter, traditional analytical approaches to problem-solving fall short when faced with unmeasurable and multifaceted challenges. These conditions, characterized by their human dimensions, extreme difficulty, uncertainty, and interconnectedness, require an alternative framework to navigate their complexity.

In response, the three principles of design engagement serve as guideposts for navigating the unknown. Through a contextual, polymathic, and evolutionary process, problem-solving can be advanced by a combination of data-driven insights and intuitive leaps, leading to the evolution and selection of actionable scenarios.

The three principles feed each other in a constant loop. To engage the dynamic nature of supercomplexity, it is necessary to rotate around the problem; as we turn and select paths forward, the process toward a solution takes on its own momentum.

"Operationalizing and optimizing are central to technology development. Design recognizes that there is much more to offer and potentially much more at stake."

Molly Wright Steenson
Historian of Artificial
Intelligence
Interview **57**

Quipu offers inclusive, fair, and accessible funding using AI-powered tools to unlock credit and capital for micro-businesses in underserved communities. MITdesignX 2019

CASE | **MICROFINANCE**

Quipu
Embracing Informality

Over 2 billion people live and work in informal conditions worldwide — amounting to approximately 20% of the world's population, expected to double to 40% by 2050 (International Labour Organization, 2023b; United Nations Statistics Division, 2023). Including gig and seasonal employees, informality already encompasses 60% of the global workforce and is on the way to becoming the defining social-economic transformation of the 21st century (International Labour Organization, 2023a). Considered by many to be impoverished and underdeveloped, these neighborhoods and their residents encompass a rich array of informal enterprises that sustain communities — examples ranging from single individuals who trade handmade products (such as clothing or bread) for other necessities to larger-scale construction, waste collection, and mobility enterprises. Cut off from access to capital, expertise, and legal protections that secure the formal economy, such ventures are vulnerable to crime, extortion, and unfair practices that limit their growth, income, and potential to accelerate community development. Launched in the vast informal community of Barranquilla, Colombia, in 2020, Quipu is a microfinance platform that advances the transition to informal economic entrepreneurship. Working closely with local micro-business owners to co-create a vision for their technological platform, Quipu's founders designed a platform to facilitate

easy, fair, and alternative capital flows to sectors historically marginalized by traditional economic systems.

Quipu is developing the first digital bank that utilizes AI and alternative data to assess the creditworthiness of microenterprises. Their approach evaluates the entrepreneur's payment capacity as well as the business's potential, and leverages Colombia's rapidly growing smartphone ownership—currently over 70% and projected to reach 97% by 2029 (Statista, 2025). Users can apply via their smartphone, receiving working capital in less than three days. Currently, the Quipu platform serves 22,000 micro-businesses and has placed over 8,000 loans of USD $200 on average. Sixty-nine percent of its clients renew the credit; 60% are women. Proving the reliability of the informal economy has attracted mainstream partners, such as Bancolombia and Claro Pay, a service of Colombia's largest mobile phone company, bringing access to millions of users and a commitment to underwrite Quipu loans up to $500,000.

As populations increase and human habitat declines, formal economic and social systems appear less and less able to meet the basic needs of non-wealthy communities. Less encumbered by outdated standards and rules, informality can be significantly more entrepreneurial in providing for those needs.

87

4 Flow With Complexity

When the tides of supercomplexity rise, we cannot command the waves — but we can learn to move with them, to chart a course through the unknown. Not by resisting the current, but by embracing its flow, we transform uncertainty into motion and vision into possibility.

"Design is free.
Design creates
as it goes."

Masayo Ave
Designer, sensory
experience
Interview **4**

The engagement principles outlined in Chapter 3 serve as touchpoints as we navigate the turbulent waters of supercomplexity. We may not be able to measure the extent of the depth beneath us or predict the speed of the current; we cannot control the weather or the finely tuned ecology of our environment. But we can envision an alternate future and move in that direction — working with, rather than against, dynamic conditions that are beyond our full understanding.

By surrendering to the fact that we cannot fight supercomplexity, in design we find an alternative source of agency by learning to "swim" within the flow (see Figure 4.1). But what does that mean in practice?

INNOVATION CONTEXT

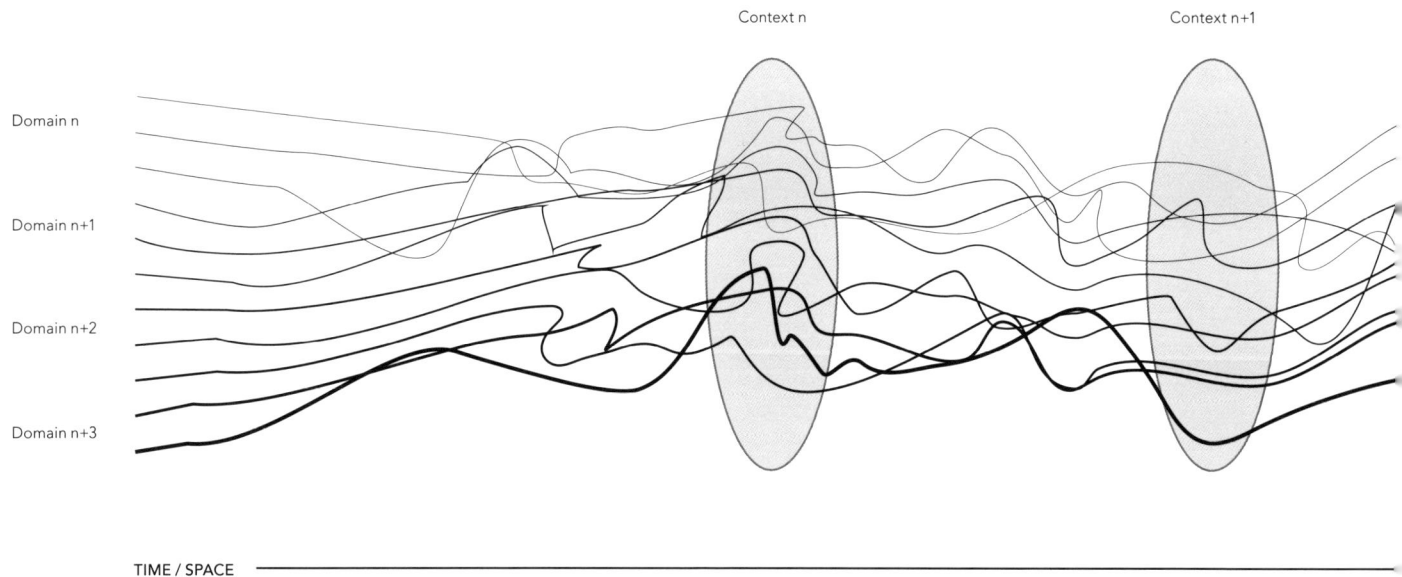

Figure 4.1 | Innovation Context. This diagram illustrates how innovators must navigate fluid, rapidly changing context shaped by shifts in technology, human needs, nature, and economic domains. What works in one place may not work in another, so it is important to navigate change across time and space to achieve solutions that are both sustainable and locally appropriate. For example, solar-powered cold storage might succeed in coastal Kenya, where sunlight is reliable and farmers often work together. But the same model could struggle in northern India, where climate patterns vary and land ownership is more fragmented. Time matters too: off-grid solar solutions that are possible in 2025 would have been unthinkable before the 1970s. Therefore, understanding the technological readiness, local governance, climate, and cultural practices is essential for achieving outcomes that are balanced and appropriate to their context.

Enter the Vortex

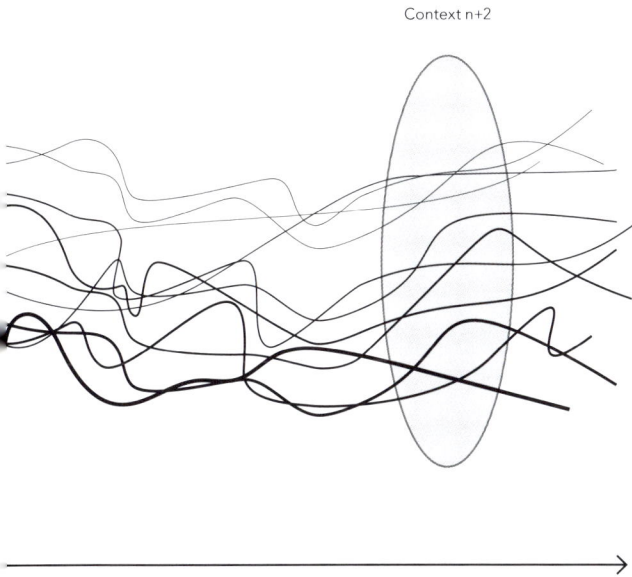

Context n+2

We describe the inner workings of design by invoking a metaphor cited in recent literature of science: a whirlpool or vortex. Ever changing, and therefore more like an event than an object, a vortex emerges from the flow of energy in a fluid medium — water, air, lava, space-time, or a stream of human consciousness (see Figure 4.2). As such, the metaphor has been used by neuroscientists to illustrate how sensory information is processed to generate perception (Freeman, 2009).

The cycles of the vortex result from an opposing force that changes the direction and momentum of the flow — like a rock in a stream, or the introduction of an opposing concept to our commonly accepted current of experience (understood as reality). The challenge is to harness the flow, reframing apparent obstructions as leverage points that serve to shape the process of onward motion.

The vortex isn't only how we stay "above water." It's also how we synthesize desirable outcomes. At the same time, it's worth noting that the process is not fully subject to control and intention. A vortex is a self-generating phenomenon in which each movement is impelled by what precedes it, while generating the next in turn. Likewise, the design process "creates as it goes;" the vortex is continually renewed for as long as a source of energy — in this case, the input of ideas and information — is present.

Direction Becomes Form

"Flowing is a dynamic balance between control and chaos — a state of continual adaptation, connection, and potential."

Gianandrea Giacomo
Psychologist, psychotherapist, and consultant

"Performance cannot be reduced to a single variable. In the extremes, it's not just strength or skill that counts — it's how you navigate the unknown."

Pietro Trabucchi
Olympic team psychologist, resilience and motivation
Interview **59**

In this chapter, we map the nature of the design process, methods of synthesis used to induce and sustain it, and how desirable outcomes emerge.

Design praxis — the iterative and collaborative process of synthesizing new outcomes — requires that we apply the full scope of our innate capabilities as problem-solvers. Flowing with complexity combines intentional and strategic action with tacit knowledge and trained instinct. The term was coined by our colleague Gianandrea Giacoma as a "dynamic process of continuous evolution and adaptation". Like the practiced swimmer whose body moves unconsciously through the medium of water to reach an intended goal, mastery of design involves a synthesis of factual information, physical sensations of the hand and mind, and receptivity to intuition.

Figure 4.2 illustrates the flow of design thought and action as a vortex (descending spiral), in which the problem–solution space continuously contracts until its two aspects are coincident at the base.

Along the way, constraints (methodologies) are employed in three discernible "cycles" — Discovery, Framing, and Emergence — that serve to synthesize facts, ideas, and options into a way forward (see Figure 4.3).

The three cycles descend at an increasing speed. Conditions at the top of the vortex, where activity is less constrained and movement is slower, are different from conditions toward the bottom, where activity is highly concentrated and moves fast. Many designers will have experienced this acceleration of activity and organization as they move toward a solution.

Each stage of synthesis can be framed as a distinct cycle, and for ease of understanding, we have outlined the cycles as a sequential trajectory. However, they function as interdependent processes in design praxis, seamlessly connected and even occurring simultaneously.

This corresponds to what we have observed throughout our research for the book. Designers often work up and down the vortex, enter at different points, or make leaps between cycles; for example, working simultaneously on the discovery of broad concepts while refining minute details. The same territory can be traversed several times over, applying different methods of synthesis and new knowledge accumulated in previous stages. The process can be compared to that of a film director, who cycles through "takes" of a scene until the vision of the director and the performance of the actors are synthesized into a new whole.

In design, the future that emerges is both predicted and produced by the form of the flow; as we cycle faster and faster down the curves of the vortex, the evolving vision and emerging solution compress to a single point. The characteristics of each cycle are illustrated in Figure 4.3 (diagrams A, B, and C).

"Design thrives in the messy, complex, and open-ended. It embraces chaos, uncertainty, and ambiguity. The path forward isn't a straight line; it zigs and zags, weaving through the unknown."

Dava Newman
Aerospace biomedical engineer
Interview **46**

Discovery. When a problem with existing life conditions becomes apparent, the design process — the vortex — is initiated by envisioning an alternative future state in which the problem does not exist. This disturbs the flow of what is accepted as everyday reality, simultaneously opening what we term the "problem–solution space": a defined conceptual zone where problem-solvers can prospect for facts, ideas, and clues from multiple areas of knowledge that could inform a potential solution. The vortex is a natural form of resolving conflicting energy flows in any medium, whether water, air, galaxies, space-time, or streams of consciousness.

Framing. The problem–solution space is continually compressed and refined as we work down the different cycles. Information and ideas gained during the Discovery cycle are framed into stories, each a cluster of related information and events that illuminate an alternate path to move from present reality to the desired outcome of the vision (or vice versa). Stories also serve as prompts for speculation, discussion, and decisions on how to proceed, thereby activating turning points in the process.

Emergence. The most promising pathways to the vision are selected by extracting and combining elements of several stories or inventing new options to overcome obstacles. As the velocity of the vortex intensifies with each concentric turn, elements increasingly fall into place until a preferred solution emerges.

VORTEX EVENT

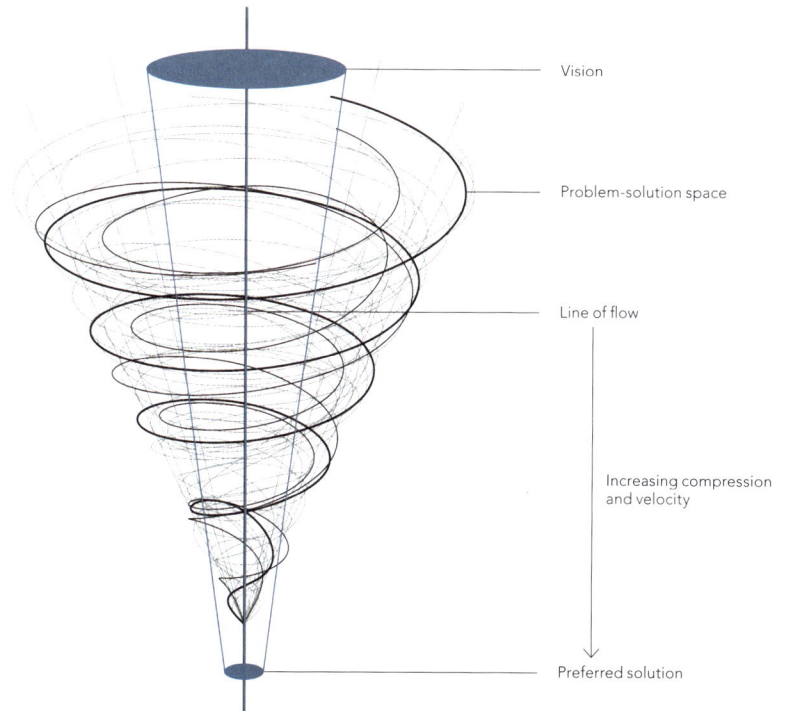

Figure 4.2 | Vortex Event. A natural form produced when flow within a fluid medium is interrupted by an obstacle or a competing force. Left: A vortex in water. Right: Dynamics of the form, which compresses and moves faster until competing forces are resolved at the bottom. In design, the vortex revolves around a 'line of flow' that connects an opening vision with the ultimate solution.

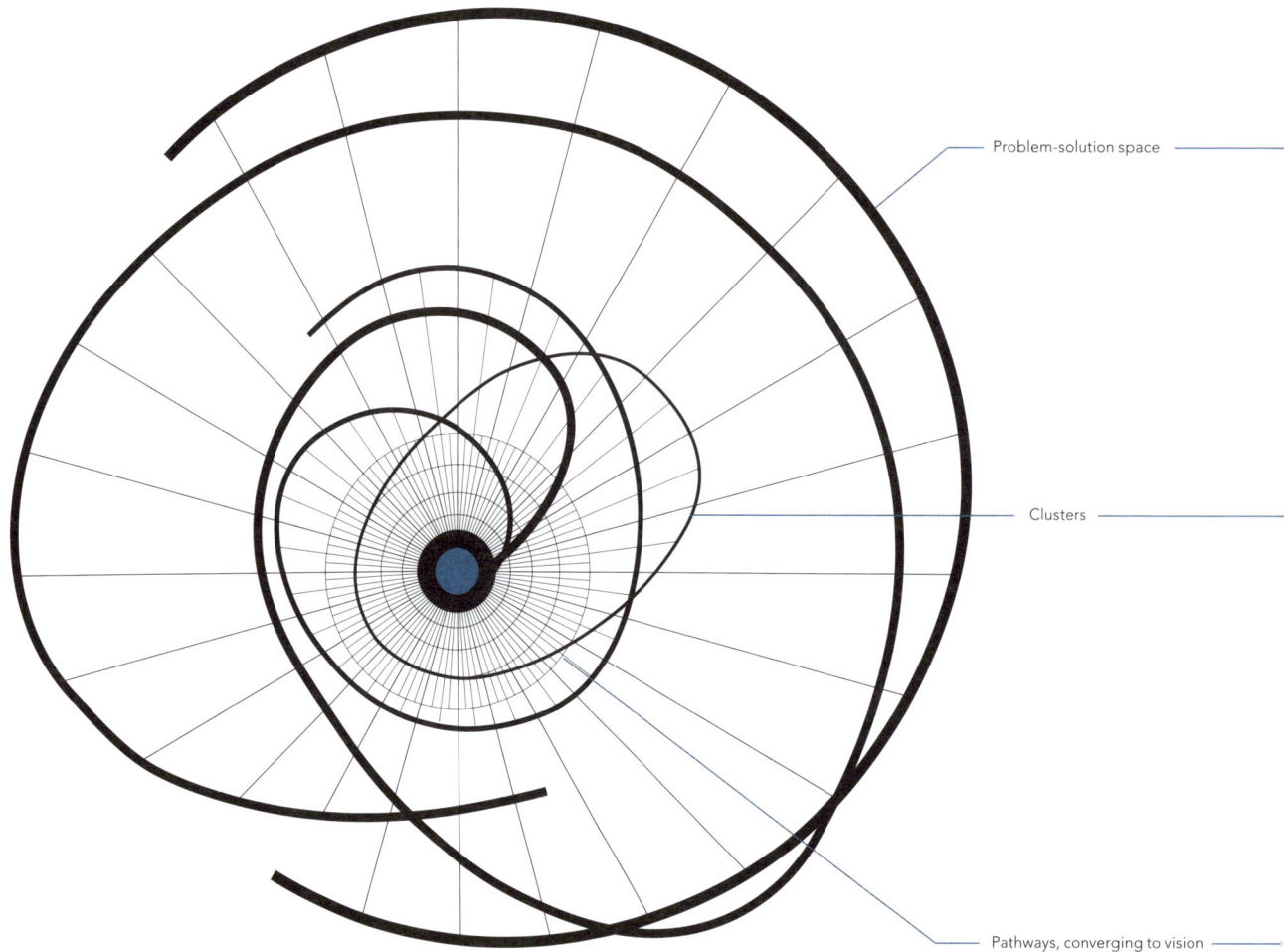

Figure 4.3 | Vortex, Cycles of Synthesis. In the design process three discernible 'cycles of synthesis' are superimposed. Left: diagram illustrates the cycles as both distinct and interconnected events, enabling designers to leap from one to the other, or to pursue alternative approaches at the same time. Shaped by different methodologies, the cycles become more focused and interconnected as we move down the vortex towards the envisioned future, or preferred solution(s) Right: Illustrates the individual integrity and characteristics of Discovery, Framing, and Emergence cycles. The design of the illustration was inspired in part by a drawing by George Courtauld.

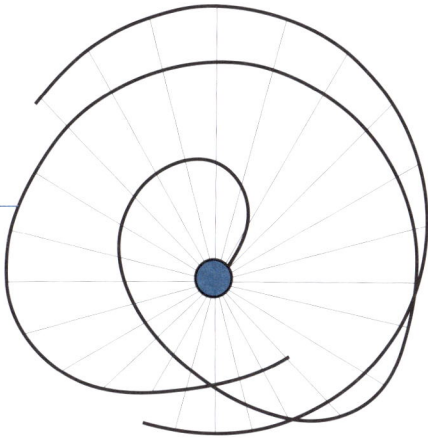

A | Discovery. The vortex first takes form when a contrasting or new concept — an alternative vision or state of being — interrupts the flow of generally accepted thinking (or our perception of reality). Thus, the primary role of a vision is to disturb the flow of everyday reality, initiating a vortex in the mind. This discrepancy shapes what we term the problem–solution space, a zone for envisioning future scenarios and prospecting facts and perspectives from multiple areas of knowledge, allowing elements of a potential solution to combine and cohere. The space is continually compressed and refined as we work down the different cycles of the vortex.

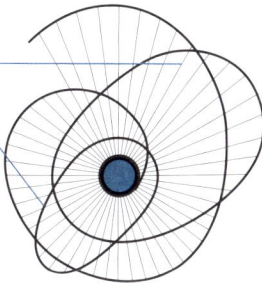

B | Framing. Information and ideas gained during the Discovery cycle are synthesized into stories, each a cluster of related events that illuminate an alternate path to move from present reality to the desired outcome of the vision. Stories also serve as prompts for speculation, discussion, and decisions on how to proceed, thereby activating turning points in the process.

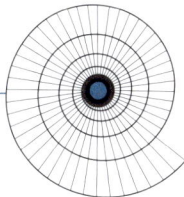

C | Emergence. The most promising pathways to the vision are selected, extracting and combining elements of several stories, or inventing new options to overcome obstacles. As the velocity of the vortex intensifies with each concentric turn, elements increasingly fall into place until a preferred solution emerges.

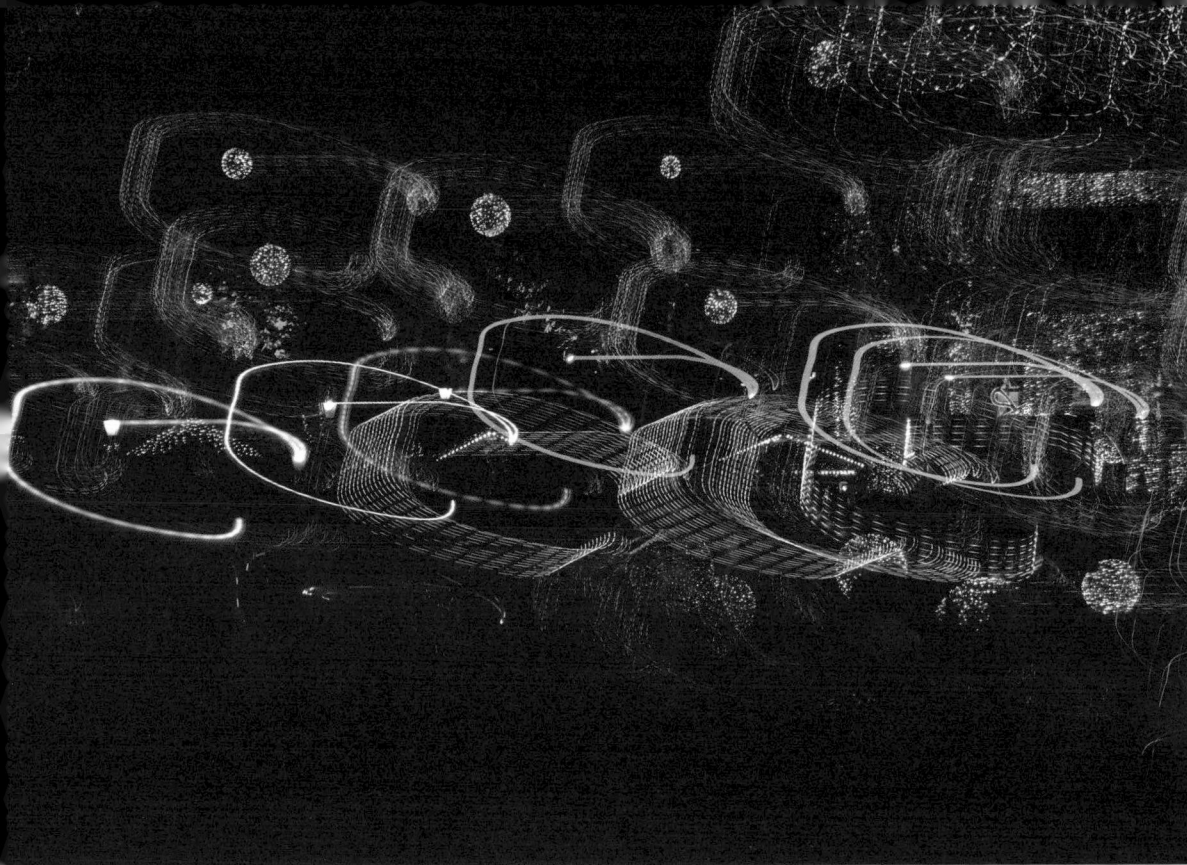

Cycles of Synthesis

Cycles of synthesis and their associated methods are explained in more detail below. The methods will be familiar to experienced designers, who typically take them for granted. Nevertheless, when working through the design vortex, they stand out as critical techniques to constrain choices and focus the process. We present the vortex as an ideal, not a prescription to follow, recognizing that the nature and application of such methods will vary by designer and project.

Discovery Cycle

METHOD | ENVISIONING

To envision means to imagine, describe, or form a mental image of an alternative state that has not yet arisen. It involves the conceptualization of possibilities, scenarios, or desired outcomes. The vision is not confined to images and may take many different forms, including words, numbers, or drawings.

The act of envisioning is often catalyzed by undesirable conditions in the present, which must be identified, stated, and interrogated as the design process moves forward. As opposed to a solution that isolates and anchors the problem in its current context, the vision can be understood as a desirable future state in which the problem does not exist (Gabor, 1963).

The vortex arises when a problem becomes apparent in the present conditions of life. This may be revealed through a formal analysis ("dimensioning the problem"), media coverage, or pushback by individuals or groups who are negatively affected. Designers set the vortex in motion by projecting an alternative vision, drawing upon a combination of imagination and practical wisdom. The vision may include fiction (giving form to an unknown future) as well as fact (based on aspects of present-day reality that are likely to endure).

"Design is the art of dreaming."

Kazuo Kawasaki
Industrial designer
Interview **33**

"To create good design, you need a culture of belonging to the previous reality, which, if known, opens infinite scenarios."

Antonio Macchi Cassia
Designer
Interview **39**

Like a rock in a stream, the vision of an alternative state acts to disturb the flow of everyday accepted reality while simultaneously opening and defining the problem–solution space.

At this early stage, the present feasibility of the vision is less important than opening a problem–solution space in which to scan for clues to the problem and opportunities to resolve it. Avoiding preconceptions, the vision is not a definition of *how* the problem will be solved, but a multidimensional representation of what life would be like in the absence of the problem.

It is easier to limit the search for a solution to known elements that we can control. However, we can only make room for innovation by allowing the problem–solution space to extend beyond what is presently considered possible.

To fail to include the vision of a desirable future state within decision-making processes would mean that problem-solvers are limited to inching forward from the known present, rather than daring to consider emergent alternatives. This would be to forgo an important factor in the way circumstances evolve. The future will tend to move toward any compelling vision, irrespective of content or feasibility in the present, with a force proportional to the power of the media

"It has to do with being able to organize a cadence of moments where you are broader in thinking and you're looking for new ideas and perspectives..."

Melissa Marsh
Workplace strategist, change management
Interview **41**

that convey it (Frenchman, 1995). This in part explains the relative power of certain visions to affect the future — exemplified by the 1939 GM Futurama described in Chapter 3, in which the vision unfolded through a simulated airplane trip across the U.S. in 1960 — 20 years hence — a vision so powerful it influenced U.S. housing, urban renewal, and interstate highway policies enacted after World War II.

Although it extends beyond the limits of present knowledge, the scope of the vision should be sufficiently contained to provide a conceptual structure for moving forward. This structure is secured by keeping the essential intent of the vision in focus throughout the process, thereby propelling designers through the vortex. As author Greg McKeown observes: "A powerful essential intent inspires people partially because it is concrete enough to answer the question, 'How will we know when we have succeeded?'" (McKeown, 2014).

Clearly defined human, environmental, societal, and financial outcomes are necessary for creating clear parameters for each stage of the process, including the later selection of proposed scenarios and options for preferred solutions.

In this way, the vision is not only the inspiration but also the objective of the process: the beginning as well as the end.

METHOD | PROSPECTING

With the vision in place, the prospecting stage serves to deepen an understanding of the context from diverse perspectives as well as possible approaches to a solution. To prospect for information in the presence of super–complexity, designers (in the manner of detectives searching for clues) must cast a wide net covering diverse fields of knowledge, gathering information ranging from facts and opinions to processes and projects — seemingly disparate "nuggets" that resonate with the problem or its envisioned solution.

Prospecting is facilitated by the vision that illuminates the territory of the search, and a polymathic design team that brings diverse understandings and perspectives to bear. This also highlights the importance of including stakeholders on the design team from the start of the process, ensuring that their ideas and perspectives are incorporated into the vision and subsequent synthesis to achieve it.

Discovery can extend throughout the design process. It initially provides a framework to identify the vision and open the problem-solution space in which to prospect for information about the problem and potential solutions. Further down the vortex, the vision that sparked discovery serves as a point of reference to assess and select among multiple options as the preferred solution emerges.

Framing Cycle

METHOD | CLUSTERING

This level of synthesis involves sorting the collected nuggets of information, moving from seemingly unrelated facts and intuitions to unified ideas and concepts. These groupings may be defined in words, sketches, diagrams, or discussions. Framed in this way, they resonate with the vision and can shine light on pathways to achieving it.

Elements of a cluster may be explicitly or implicitly related to one another in various ways; for example, in appearance, spirit, pattern, language, viewpoint, form, or other commonalities. The art of framing is to make clusters of roughly the same metaphorical weight or significance to the problem and its solution. This is a subjective judgment.

Some clusters may include clues found across multiple fields in the problem—solution space, whether science, economics, sociology, or philosophy. Other clusters may draw upon just two or three fields, or even one, represented by a single nugget of information. A single fact or idea from one field may be so powerful that no pathway to a preferred solution can be envisioned without taking it into account, and so it is framed as its own cluster. Other clusters that frame elements from two or more fields can have equal potential to be part of a successful solution.

"Design is facing problems not only of a technical or functional nature, but also of meaning. Therefore, framing, envisioning and solving cannot be separated."

Gloria Barcellini
Architect
Interview **6**

How many clusters should we make? As a method of synthesis, we suggest the "Frenchman Rule" of limiting the number of clusters to 6 +/– 1, no matter how many elements are available to work with.

Six is a number we can comfortably hold in our minds; more than seven is too many to deal with at once, and less than five is too few to assure diversity. The rule is consistent with psychological studies, which have demonstrated that working memory caps at around seven items (hence, most telephone numbers are seven digits long (Miller, 1956).

The rule applies to any complex design problem and collection of data and ideas. If there are more than seven clusters, some need to be combined or discarded; less than five, some should be broken apart. The rule helps to ensure that clusters are balanced, manageable, and well-formed while being sufficiently diverse to provide a foundation for further synthesis.

In the detective work of prospecting and grouping nuggets of information, there comes a point when a cluster, or thread of evidence, is ready to be framed as a convincing story of the case. In this way, the efficacy of the various clusters is tested and strengthened. Certain groupings are discarded, and others are selected to move forward.

"Framing is solving. Solving is also framing, and discovery. Design discovers through interactions."

'Kayode Agbalajobi
Real estate asset manager
Interview **1**

"A product [solution] is the synthesis of a series of 'stories'."

Anty Pansera
Art and Design historian
Interview **50**

"Design, in its simplest
form, is compelling
storytelling. If you're a
great designer, you're a
great listener because
you are echoing
thousands of stories."

Andrea Chegut
Innovator, research scientist, and
financial econometrician
Interview **16**

METHOD | STORIFYING

The everlasting appeal of myths, oral storytelling, and
epic sagas reveals the power of stories to humanize
and engage the unknown; to witness the facts of life
recombined into a fiction that helps us to make sense of
the world in which we find ourselves.

Epic narratives in particular provide us with a framework
and trajectory to understand how we might map a path
from vision to action, goal to attainment. Archetypal texts
such as *The Epic of Gilgamesh* and Homer's *Odyssey*
follow the protagonist as they invent their way over
seemingly insurmountable obstacles to reach a projected
destination. Propelled by the momentum of the journey,
we travel with them as their pathway emerges.

Storifying can be understood as the process of
synthesizing a narrative from the facts and ideas at
hand — in this case, illuminating routes to the vision by
recasting prospected information into plotlines or
journeys of sequential challenges that can be tackled
individually.

Stories are also a medium of weaving fact and fiction
to generate a more robust view of the road ahead. The
only requirement is that the plot flow from beginning
to end within a specific physical and social context,
thereby connecting actors, events, and outcomes.

By framing clusters as stories, continuity gaps and
inconsistencies become immediately apparent, and
information is made more emotionally relatable and
memorable. This helps to cohere information and
communicate complex or unfamiliar ideas, facilitating
engagement by capturing the attention of target
audiences.

Furthermore, the integration of context will radically
impact the way a story (or pathway to a solution) is

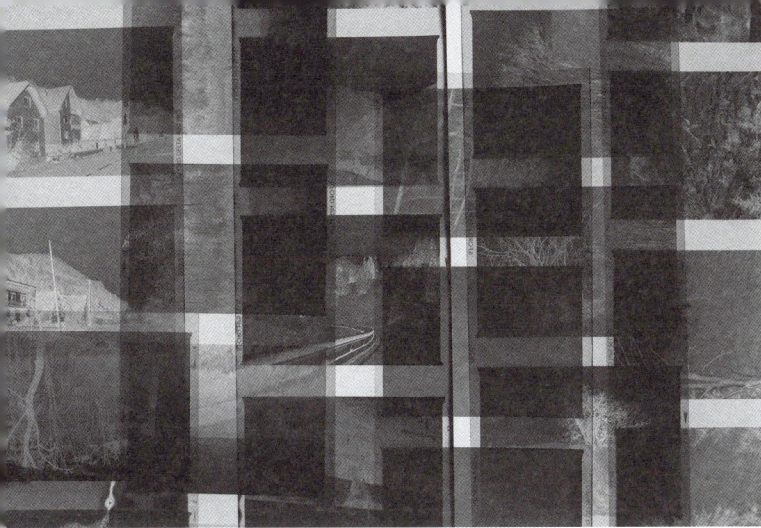

"A change in language can serve to change reality. Changing the way we talk about an issue is already, in a way, a redefinition of the situation."

Rubén Fernández-Costa
O'Dogherty
Writer and journalist,
information sciences
Interview **26**

received. The efficacy of a story will depend on the extent to which design team members resonate with the narrative and desire to follow where it leads. In this way, storytelling is always an act of co-creation. In particular, stakeholders — as readers, interpreters, and meaning-makers — do not passively receive the material of the story; they mythologize within it and carry its message and relevance forward into the context of their own lives (Frenchman, 1990).

It is worth noting that stories have always been part of the design process, integral to any method that serves to illuminate or explain the rationale of a design. In recent years, narrative techniques have been amplified by advanced tools and methods that incorporate digital media. For example, we can instantly generate views of a story in context by using artificial intelligence (AI).

Despite the importance of narrative in motivating transformations, it has only been acceptable as part of design discourse in educational settings within the past decade or so. Prior pedagogy, adhering to the modernist mantra of "less is more," dismissed the need for narrative explanation. It was believed that form and form-making should exist independently of narrative content or expression. Arguably, this paradigm led to generations of soulless buildings, barren public spaces, and repetitive infrastructure bearing no relation to history, culture, or context — thus communicating little and failing to become a part of the collective meaning-making that is integral to a structure's value over time.

Framing involves searching, selecting, and evaluating prospective data points by creating 6 +/– clusters, to constrain and manage the process of synthesis. We recast the known and speculative information framed by each cluster into a story of how we move from the present to a future state in which the problem does not exist. These stories provide the foundation for the next cycle of synthesis.

Emergence Cycle

METHOD | PATHFINDING

In the cycle of Emergence, stories are the raw material for enabling a higher level of synthesis. Much of the foundational work of synthesis has already taken place in the process of troubleshooting the elements of a story to ensure they work together as a synthesized whole — organizing disparate elements into narrative threads and thereby easing the task of finding a path forward.

Stories are also inherently adaptable. In the manner of CRISPR gene editing technology (Doudna & Charpentier, 2014), we can snip elements of existing stories and rearrange or replace them to achieve different outcomes, providing that logical consistency is preserved.

"In the design process, you have to start at the top and then arrive at a single, very powerful message or solution. It's also a process of unifying. And in this process I think the most important thing is to be able to discard options and to know how to say "no" to an idea."

Rubén Fernández-Costa O'Dogherty
Writer and journalist, information sciences
Interview **26**

"Design is an iterative process. We find ways where we can improve, do it better, or do it differently, and then we try again. Sometimes we demolish, sometimes we repurpose. The act of doing and solving is not the end of the road. It's yet another discovery which starts the process all over."

'Kayode Agbalajobi
Real estate asset manager
Interview **1**

The next step of pathfinding involves pursuing a shortlist of the most promising storylines. Here we suggest limiting the shortlist to three. In practice, designers often find themselves landing on a triad of alternatives: two extremes and one in the middle. These extremes may, for example, relate to cost, levels of risk, or impact — who benefits or loses?

The chosen stories should embody all the elements of a well-rounded whole, including all the necessary factors for a desirable solution, even if not yet fully developed. As such, they may be considered "proto-solutions" — the prequels or early life forms of an emergent solution that will be selected and implemented.

Ideally, proto-solutions offer distinctly different directions forward. The paths are not predetermined but evolve as challenges are encountered, options emerge, and further choices are made.

METHOD | SELECTING

A pathway starts to take form as proto-solutions are eliminated one by one on the basis of whether or not they align with the essential intent of the vision.
There proceeds a process of negotiation and selection to determine which proto-solution is preferred, or whether a synthesis can be made of the most appealing features in each — for example, lower cost and higher benefits to those who need them most.

This process is accelerated by the form of the vortex, in which the problem–solution space is compressed as further constraints are applied or challenges uncovered. To use the terminology of complexity theory, the challenges we encounter on the path to a solution can also be understood as leverage points: sites of testing and decision-making in which a seemingly minor choice between two directions can have significant impacts on the eventual outcome (Meadows, 2012).

Strategies to overcome challenges, or leverage an advantage from obstacles, include:

Backtrack to prospect for missing information. For example, it can be helpful to find evidence of how a similar challenge has been engaged elsewhere (whether or not successfully). However, it is important to be aware that there may be few precedents for meeting the challenges of a supercomplex problem.

Redefine the problem–solution space. This implies adjusting the vision, which may improve the potential fit of one or more proto-solutions, making challenges easier to overcome or opening alternative pathways that bypass the challenge entirely. This method is often dismissed because fundamental decisions may need to be reconsidered (potentially delaying the process).

Invent a novel way to address the issue. This is the riskiest approach to a challenge, but it can leverage significant benefits. Casting the problem within the framework of a story facilitates invention by highlighting missing links in the narrative and thereby the credibility of a potential solution.

"Design is a process of seeing the world as reconfigurable. A synthetic approach that re-configures arrangements of people, matter, and combinations of them to systematically produce better outcomes. But it is distinct from optimization. Optimization is not necessarily a creative act and design always is."

Nicholas de Monchaux
Designer, author, educator
Interview **24**

"Anything wrong with my design reflects my problem frame — a beautifully honest reminder that the way we see the world defines the challenges we take on."

Tomoko Azumi
Interior designer, products, furniture and space
Interview **5**

Designers gain a deeper understanding of the problem's context and impact as elements are tested and new threads of synthesis emerge. Any gaps or disjointed information signal the need for more knowledge or a new approach. Testing can take different forms to better understand the strengths and weaknesses of various choices, for instance: public surveys, experiments, simulations, and demos.

The key to evolving a successful solution is to test early and often at various scales, taking into account different levels of detail and context. It is important to adopt an iterative approach; re-testing occurs throughout all cycles of synthesis to uncover issues, gather feedback, adjust, and improve solution ideas in real time. This helps to align the solutions with expectations as they are formed, rather than waiting to be surprised when previously unknown issues and opposition surface in public.

The options that best align with the vision are selected to move forward, while others are discarded. In this way, the vision continues to play the role of selector as part of an evolutionary process; from the options that survive, a "preferred solution" emerges.

We call the solution "preferred" because it differs from the singular unique answer favored by analytic problem-solving. The methodologies of synthesis allow for greater nuance, recognizing that the essential intent of the orienting vision may be accessed in different ways.

Preferred solutions are courses of action that are considered best aligned with outcomes envisioned at the start of the process. The rounds of iterative testing, elimination, and innovation demonstrate that these solutions are implementable, supported by stakeholders, and will have a significant, meaningful impact on short-circuiting the problem and improving the lives of those most affected.

Emergence is an evolutionary cycle in which fitness to the vision becomes the test for naturally selecting among multiple paths forward. It involves prototyping nascent proto-solutions, interacting, testing, and iterating or refining based on feedback and constant reference to the vision. Allowing the momentum of the process to guide the path forward, designers feed further development as the preferred solution becomes increasingly apparent.

Complexity's Flow

"Design has entered the
realm of the highly abstract,
so it has become difficult
to understand. As design
shifts from physical forms to
intangible experiences, it's
becoming harder to grasp,
yet more essential than
ever."

Dai Fujiwara
Designer, creative director
Interview **29**

In this chapter, we have seen how the vortex of design involves multiple methods of synthesis that actively engage supercomplex problems. Each cycle takes on the entirety of the problem from a holistic and polymathic point of view, compressing the problem–solution space by retaining elements that support the vision while reframing, removing, or replacing those that do not.

New ideas and paths to a solution may occur at any point in the process, which must be sufficiently flexible to allow for continuous learning and realignment. This requires a mindset ready to embrace uncertainty and remain open to new perspectives, facts, and intuitions; a state of resilience and receptivity crucial for responding to the unexpected challenges of rapidly changing circumstances.

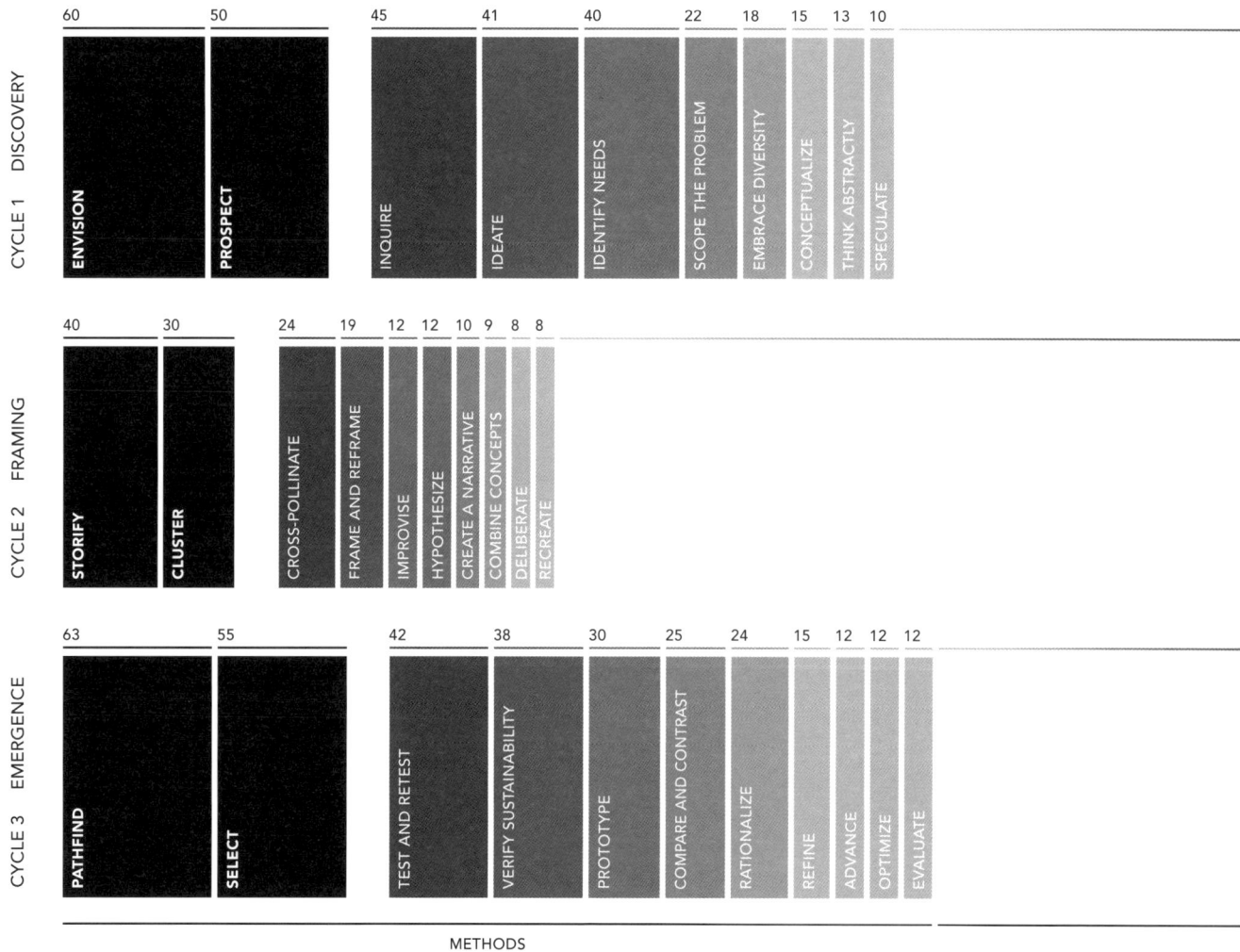

Figure 4.4 | Methods of Synthesis. Data analysis of interview transcripts revealed key methods applied in each of the cycles: Envision and Prospecting (discovery cycle); Storify and Cluster (framing cycle); Pathfinding and Selecting (emergence cycle). Numbers in the graph reflect ranking of concepts based on their frequency and contextual significance across interviews for this book (see Appendix, Methods).

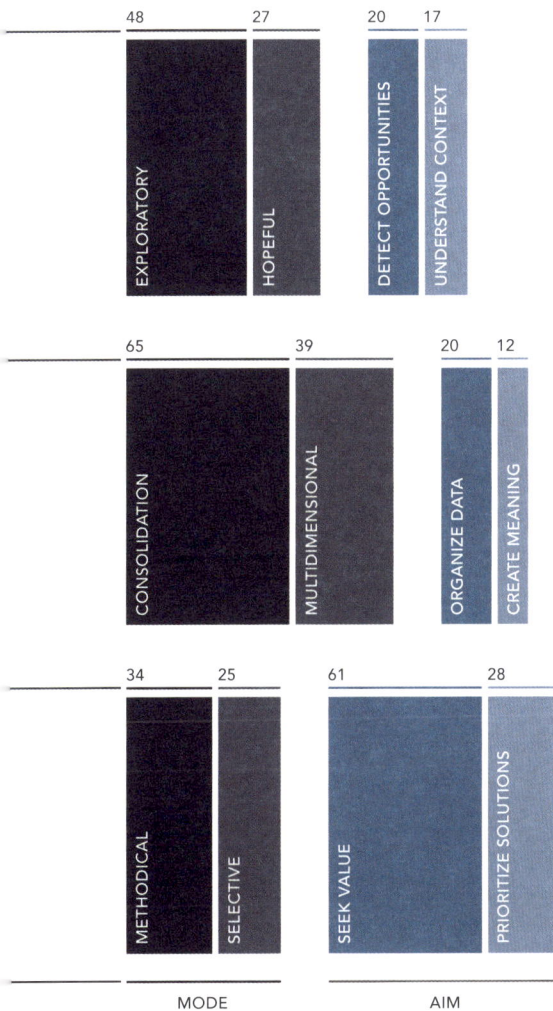

At root, the design mindset is that of the realist — to act, relate, and make in a world where change is the only constant. How do designers sustain the energy to engage the world in this way? The key is hidden at the center of the vortex. As designers navigate the flux of supercomplexity through a strategic combination of intention and intuition, there is a feeling of moving ever closer to a solution; both the "source" and final point of orientation that is simultaneously discovered and created.

Indeed, as the vortex reaches its greatest velocity during the Emergence cycle, designers often sense that the evolving solution is pulling them forward. It becomes increasingly clear which paths to abandon and which to test and strengthen. In other words, the conscious streamlining of the process cedes to the flow and inner momentum of the vortex itself.

Chart — Modes and Aims

	MODE			AIM	
Cycle 1	EXPLORATORY (48)	HOPEFUL (27)		DETECT OPPORTUNITIES (20)	UNDERSTAND CONTEXT (17)
Cycle 2	CONSOLIDATION (65)	MULTIDIMENSIONAL (39)		ORGANIZE DATA (20)	CREATE MEANING (12)
Cycle 3	METHODICAL (34)	SELECTIVE (25)		SEEK VALUE (61)	PRIORITIZE SOLUTIONS (28)

Modes and Aims. Data analysis of interview transcripts revealed key modes and aims endemic to each cycle. Cycle 1: The mode is exploratory and hopeful, with the aim of understanding the context and uncovering latent opportunities; Cycle 2: The mode is consolidative and multidimensional, and the aim is to organize the data to create meaning; Cycle 3: The mode is more directed and selective with the aim of seeking preferred solution(s) with the greatest value for all stakeholders. Numbers in the graph reflect ranking of concepts based on their frequency and contextual significance across interviews for this book (see Appendix, Methods).

Solid Ground

"Good design has two
aspects: codes that
are objective… and
narratives that provide
meaning."

Luciano Galimberti
Designer
Interview **30**

As we exit the vortex and progress to the grounding of
the next chapter, it's worth pausing to reflect on flow as a
natural phenomenon.

The word "magic" has too often been applied to the
design process, idealizing creativity and obfuscating the
inherent pragmatism and rigor of design. As a result, the
design skill set has conventionally been overlooked when it
comes to real-world problem-solving, instead being viewed
as an aesthetic add-on or a fashionable marketing strategy
to enhance the appeal of a brand.

However, if there is any apparent "magic" to the design
process, perhaps this is it: not the wild creative leap, but
instead the fluid mechanics of the process as a vortex,
carrying the designer forward in its flow toward the
eventual emergence of the X — the transformation from
a collection of segmented parts to a greater whole. In its
innumerable interconnections, a tensile structure that has
been strengthened at every turn, the resulting solution has
a ripple effect that exceeds the original vision.

It is at this point that the myth of magic ceases to be
applicable. What we are witnessing is not the mysterious
art of "making happen," but simply what it looks
like to apply the full scope of our human capabilities
through cycles of synthesis, acting in relation to the
supercomplexity of a turning world.

Roofscapes transforms unused rooftops into green spaces, helping cities to be cooler, greener, and more resilient.
MITdesignX 2020

CASE | **CLIMATE CHANGE**

Roofscapes
Reducing the Heat Island Effect

Projections indicate that by 2050, a majority of the world's cities will be too hot to safely host the Summer Olympic Games (Robinson & Dewan, 2024). The intensification of urban heat islands linked to global warming became apparent during the 2024 Paris Olympics, when spectators and athletes struggled with scorching temperatures that threatened their health and performance.

The primary source of the problem can be seen from a tall building in any city. Below lies a wasteland of rooftops, hostile to humans, unvisited by wildlife, covered in materials that absorb sunlight, which then radiate heat down into structures and streets beneath.

The problem is particularly acute in Paris, due to the city's relatively low amount of green space, dense stock of historic masonry buildings, and zinc roofs that readily absorb solar radiation, causing roof temperatures to soar as high as 194°F, or about 90°C, on summer days (Voiland, 2024). Studies have shown that Parisians face the highest risk of dying from heat waves compared to other European cities, across all age groups (Gasparrini et al., 2017; Masselot et al., 2023).

Committed to mitigating the catastrophic impacts of climate change on cities, architect-developers Olivier Faber, Eytan Levi, and Tim Cousin founded Roofscapes Studio to design and deploy a universal modular system of interconnected rooftop gardens and communal spaces. These are designed to shade and cool roofs while increasing the social and financial value of the buildings on which they rest. During a demonstration on the rooftop of a former town hall in Paris, the system lowered the interior temperature of the building's top floor by 30.6°F (17°C) on a hot summer day (Gattupalli, 2024).. Working hand-in-hand with the City of Paris, Roofscapes Studio is now collaborating on the deployment of its solutions on several other building types, such as public high schools, private condominiums, and office buildings.

As global warming advances and the problem of urban heat islands intensifies, Roofscapes offers a mass-customized solution that reimagines underutilized roof surfaces, transforming them from overheated wastelands into cooled, shared outdoor green spaces.

Blue Lagoon
Playing Chess with Nature

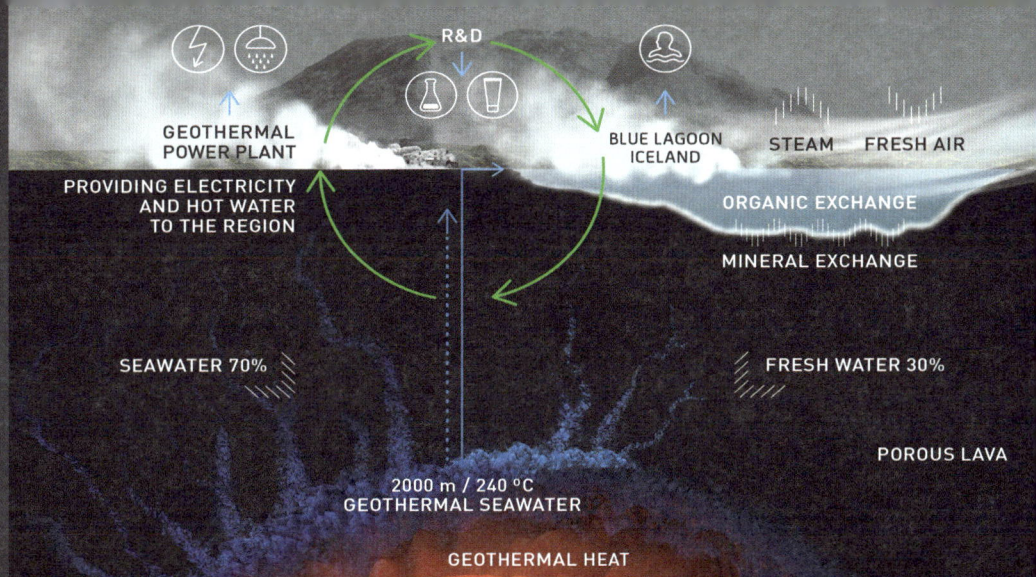

The Blue Lagoon exemplifies the principles and processes advocated in *Designing the X*

META CASE | **DESIGNING IN THE FACE OF SUPERCOMPLEXITY**

The story of the Blue Lagoon is an alchemical paradox. Water heated deep within the earth becomes light and power; industrial effluent becomes a source of beauty and healing; a wasteland next to a geothermal power plant becomes a place of pilgrimage for travelers in search of wonder and well-being.

For many, the images of a luminescent oasis amid the jagged rocks of a hostile lava field have become symbolic of Iceland. Situated on the Reykjanes Peninsula, the Blue Lagoon is embedded within a landscape formed as tectonic plates slowly wrench apart, releasing intense volcanic and geothermal activity. Deep within the rift, reservoirs of freshwater and seawater are mixed at high pressure and temperatures reaching 240°C (464°F). The Svartsengi power plant taps this superheated water to generate electricity and regional heating, before the comparatively cooled liquid flows into the ancient lava field surrounding the site.

Held by the porous basalt, a lagoon is formed and replenished every 40 hours. Rich in silica, minerals, and algae, the steaming water takes on a milky-blue color produced by light reflected by particles of silica, which coats everything beneath the surface with a porcelain-like white finish. Research has shown these same particles form a natural protective coating on the skin, while the presence of sulfur, calcium, magnesium, and blue-green algae serves to heal and rejuvenate.

The site has also gained the reputation of a design icon, with multiple awards celebrating its continuity of architecture and landscape. While praised for its aesthetic integrity, ecological sustainability, and exceptional success in building a globally recognized brand, the design of the Blue Lagoon — like the geological situation of the site itself — amounts to far more than what can be seen on the surface.

The lagoon is a remarkable example of how design engages supercomplex, even volatile, environments and unexpected circumstances; in this case, ranging from construction challenges, to global financial crises, to the repercussions of COVID-19, and a series of ongoing volcanic eruptions. By engaging in a flow of unpredictable forces, the Blue Lagoon inherently exemplifies the principles and process of *Designing the X*:

The design is deeply contextual, adapting to the realities of Iceland's geothermal landscape, cultural history, and changing economic conditions.

The collection of facilities and programs on the site stems from the visionary thinking of founder and medical doctor Grímur Sæmundsen. His collaboration with scientists, engineers, hospitality experts, architects, economists, cultural historians, and geologists continually shaped how challenges and opportunities were navigated—a truly polymathic approach from the start.

Far from fixed form, the design of the Blue Lagoon is evolutionary — responding over time to the interplay of natural forces, social dynamics, and shifting global events.

These principles have led to the emergence of the Blue Lagoon as a globally recognizable, iconic experience. This is the outcome — still evolving — of a design process developed by working with the facts of a changing context, sourcing knowledge and ideas from across disciplines, and being open to a process that creates as it goes.

Like a vortex repeatedly turning through cycles of synthesis, each supercomplex challenge further tightens the cycles; leading to new forms, new pathways, and new ways of preserving the paradox of the Blue Lagoon as both a unified experience and a process in flux.

Discovery

In 1997 when Grímur brought in Sigurdur Thorsteinsson, co-author of *Designing the X* and a partner at Design Group Italia (DGI) it was a meeting of minds. Two polymaths with different backgrounds but with complementary skillsets who bonded through a shared vision. Grímur saw the potential of bringing design into the control room, and together they merged business and design in a way that changed the evolution of Blue Lagoon. The original, conventional design brief grew into a comprehensive design strategy inseparable from the company's business goals.

Through prospecting the site's complex physical and social context, a vision emerged of what the Blue Lagoon could become. The lagoon had first appeared in the late 1970s, when surplus heated water from the nearby Svartsengi geothermal power plant accumulated in an ancient nearby lava field. By chance, healing properties were discovered by local residents who came there to bathe, claiming that the mineral-rich waters could cure various skin conditions. When these claims were confirmed by scientific research, the Blue Lagoon gained credibility as a health clinic and spa.

Continuous cycles of discovery and testing have further deepened the understanding of the site and its particular scope of interconnected challenges. The resulting vision intertwines physical, social, environmental, and economic well-being in relation to the extraordinary visual features of a volcanic landscape. It is a vision that speaks to how we survive and thrive as human beings, alive to the sensate immediacy of our surroundings.

Framing

Developing a 360° perspective encompassing multiple viewpoints involved a process of clustering facts and ideas from different areas of knowledge, combining Earth sciences and medicine, folklore and literature, social history and psychology. The range of sources and themes reflects the lack of a comparable precedent. The Blue Lagoon is arguably the first destination of its kind; it effectively launched the category of hot spring tourism in Iceland, uniquely placed to tap into associations between water, well-being, and human contact with the forces of nature.

Storifying these thematic clusters has been central to establishing the Blue Lagoon's success as a brand. The sublime geology of the site, the grounding in scientific research, the history of hot spring bathing, and the nation's ancient mythologies of water, life, and landscape — these are all

factors of a narrative woven throughout the visitor's experience. The presence of wood, lava, moss, and glass as part of the architectural material palette, the balance between shelter and exposure, the play of darkness and light; the scenography of the site as a whole is calibrated to evoke emotions that extend beyond a single visit, restoring our sense of wholeness with nature.

The strength of the story — which, as a physical, psychological, and communal experience, becomes a truth — has been demonstrated by the way the story has been borrowed, interpreted, and represented by others. From the use of the Blue Lagoon's photography by tourist boards and airlines to visitors' sharing of memories on social media when the site was closed during the COVID-19 pandemic, the story of the Blue Lagoon has become part of Iceland's own mythology and presentation to the world: not a fixed narrative, but an evolving saga.

Emergence

The existential threats to the Blue Lagoon over the decades have imposed their own process of natural selection, leading to new pathfinding.

Shortly after the opening of its first facility in 1999, came 9/11, and more was to follow. Shortly after Blue Lagoon doubled its facilities and was expanding its skincare line, the financial crash of 2008 caused the Icelandic króna to lose nearly half its value in a matter of weeks, seriously impacting the nation. The shockwaves triggered by this event and the personal repercussions for individuals, families, and communities cannot be understated. Then, only two years later, the massive eruption of Eyjafjallajökull sent a vast ash cloud into the atmosphere, grounding flights across Europe and serving as a powerful reminder of nature's disruptive force.

Under the leadership of Grímur, interpreting threat as opportunity, the team behind the Blue Lagoon sought to adapt the company's business strategy and visitor experience in response to changing contexts. By expanding the scope of revenue sources and reconsidering the range of hospitality and experiential offerings for travelers visiting for a day trip or an extended retreat, the structure of the company changed in ways that ultimately strengthened its ability to survive extended closure during the COVID-19 pandemic. Finally, the risks posed by unpredictable volcanic activity require evolutionary approaches that respond to the Blue Lagoon's supercomplex natural and operational environment. In March 2021, after 800 years of dormancy, the volcanic system across the Reykjanes Peninsula experienced a dramatic spike in seismic activity, setting off a series of eruptions that threatened nearby infrastructure and forced the evacuation of the fishing village of Grindavík — deeply affecting the community and its livelihood. The Blue Lagoon sits within this evolving volcanic system.

The shifting of tectonic plates and the contingency of social, economic, and ecological factors continue to shape the fate of the Blue Lagoon, causing its form — as a physical site, a business, and a recognizable experience — to remain emergent.

Success is defined not by the ultimate outcome, but by the consistent engagement of supercomplexity through the design process. By virtue of its position at the nexus of interconnected flows and forces, this is a site that will never remain static — yet the constancy and optimism of the orienting vision provide a touchstone for moving through uncertainty. Design for an unknown future will always be a process of "playing chess with nature." Played with optimistic vision and skillful collaboration, the outcomes have the potential to be beautiful, functional, resonant, and resilient.

TIMELINE

The Blue Lagoon has gone through a series of crises, but every crisis has been used to strengthen the foundation and prepare for what is next and future expansions.

1976 — The Blue Lagoon is formed by accident when surplus water from the nearby Svartsengi geothermal power plant accumulates in a lava field. Minerals in the water, including silica and sulfur, result in the lagoon's milky blue color.

1981 — A local resident, Valur Margeirsson, becomes the first person known to bathe in the lagoon, claiming it helped relieve his psoriasis. This attracted attention to the potential healing properties of the water.

1987 — The first public bathing facility is officially opened. Basic amenities such as showers and changing rooms are introduced, marking the start of the lagoon's transition into an attraction that now accommodates over 1 million visitors each year.

1987 — Official studies confirm the healing properties of the Blue Lagoon water.

1992 — The Blue Lagoon's first official research and development center is established to study the lagoon's unique mineral composition and its effects on skin conditions, particularly psoriasis.

1993 — The Blue Lagoon opens its official medical clinic.

2007 — The Blue Lagoon Clinic Hotel is opened. The Blue Lagoon geothermal bath is renovated, adding restaurants, new water experiences, and doubling in size.

2005 — The Blue Lagoon Clinic Hotel, later named Silica Hotel, is opened.

2006 — DNA molecular analysis of the Blue Lagoon water is conducted.

2003 — The Blue Lagoon official research and development and harvesting center is established to study the lagoon's mineral composition and its effects on skin conditions, particularly psoriasis.

1999 — The Blue Lagoon facilities are relocated to a newly developed site due to the expansion of the Svartsengi geothermal power plant.

1997 — A modern public bathing complex is constructed, including a restaurant, treatment center, and spa services.

1997 — Design Group Italia (DGI) is initially commissioned to design the packaging of the Blue Lagoon's skincare range — a project that became a holistic rethinking of the site, the business model, and the vision of the Blue Lagoon as a globally recognized brand.

1994 — The Blue Lagoon company is founded to manage and develop the site further, enhancing facilities and expanding services for visitors.

1995 — The Blue Lagoon launches its first treatment product line based on its unique silica mud.

2008-09 — The global economic crash leads to the strategic reframing of the company's financial strategy, later allowing the Blue Lagoon to survive loss of revenue during the COVID-19 pandemic.

2010 — The Eyjafjalla eruption pushes the company to align its business and design strategy

2012 — National Geographic selects Blue Lagoon as one of the 25 wonders of the world due to its unique origin, active elements, ecosystem, and healing power

2013 — To underline the strategic shift, a comprehensive rebranding is launched.

2014 — The construction of The Retreat as a new luxury development.

2015 — The Blue Lagoon obtains two US patents and one EU patent for the use of its unique and proprietary bio-actives, silica and micro-algae, in skincare products.

2020-21 — The BL+ skincare line is launched with the new BL+ Complex build on the patents.

2020 — The Blue Lagoon temporarily closes due to the COVID-19 pandemic, but intensifies global communications and staff-training initiatives, leading to a reopening later in the year.

2018 — The Retreat at Blue Lagoon Iceland opens, featuring a luxury hotel, a subterranean spa, and a private lagoon. This high-end development elevates the Blue Lagoon into an ambitious high-end wellness destination.

2016 — Another major expansion, adding more bathing areas and upgraded spa facilities to accommodate the growing number of visitors.

2023 — Seismic activity near the Reykjanes Peninsula results in temporary closures as a precaution, highlighting the volcanic nature of the region. The Blue Lagoon reopens after ensuring visitor safety.

2023 — The Blue Lagoon opens Highland Base in the center of Iceland, initiating its expansion into other destinations.

2024 — Further seismic activity and partial damage to the Blue Lagoon parking facilities prompts careful reframing of the site-design for renewed defenses and navigation.

2025 — The Blue Lagoon experience and arrival layout are redesigned in response to a period of seismic and volcanic activity that resulted in lava flowing over the parking and arrival center.

The next inflection — The Blue Lagoon continues to evolve in response to changing contexts – each inflection a leverage point in a process that creates as it goes.

5　The Collective and the Conductor

When all the pieces are moving fluidly together, it becomes impossible to distinguish between who leads and who is led.

Those who flow with the vortex, we call "designers" — a term that can apply to problem solvers of any discipline who are skilled at navigating the cycles of synthesis.

To flow with complexity requires both practice and expertise. Didactic learning and facts are not enough to engage supercomplexity; we need the visceral sensation of being in the flow to understand how to work with it.

Consider learning to swim. We may be taught the basic principles — how to move the arms and legs — but only the mind can coordinate and synthesize the motions into continuous flow. This is the "praxis" of swimming: practical activity toward achieving an envisioned goal. Through engagement with the sensation of moving in water, something emerges that is greater than the sum of its parts: a swimmer.

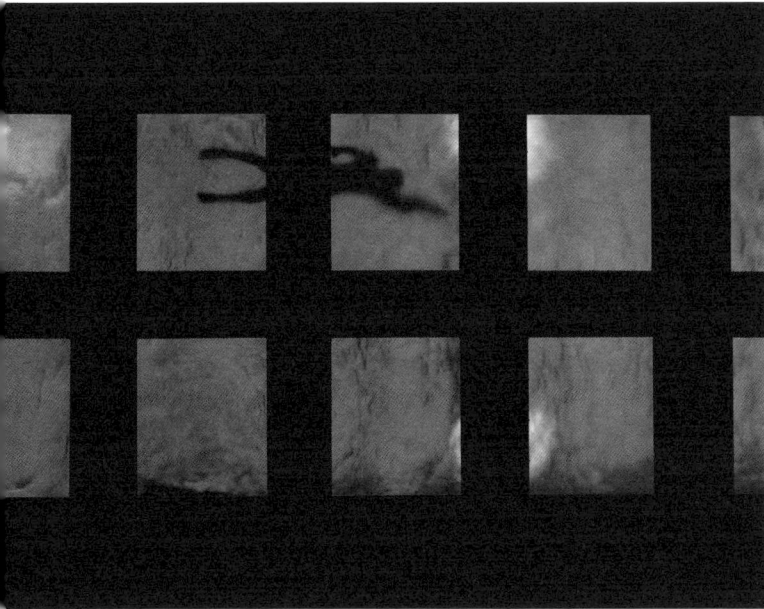

Likewise, design involves the eye, the mind, and the hand practicing in the medium to achieve mastery. The act of making (whether drawing on paper or screen) is equivalent to the feel of motion in water, where the muscles and eyes feed sensory information to the mind. Like a swimmer, our minds and bodies learn to flow with complexity, and the flow of ideas is sustained by working within a team of different skill sets and experiences.

In this chapter, we investigate what it means to design as a team, identifying the essential human qualities that uphold the process as it engages supercomplex challenges to spot innovation opportunities, as well as outlining the organizational structure conducive to processes of synthesis: the act of "making things" collectively.

"The designers who are not just solving small problems, but also presenting a big vision, are the designers who are needed now."

Masayo Ave
Designer, sensory experience
Interview **4**

Polymathic Practice

When the interconnected members of a team are working together in flow, they are no longer an assortment of distinct individuals but nodes within a dynamic network, akin to a living organism. The solutions developed by that team transcend what each member can produce alone. Together, they are a manifestation of the X in practice.

In Chapter 3, we introduced the importance of a polymathic approach. Working with supercomplexity demands knowledge of multiple fields, necessitating the involvement of specialists and stakeholders with divergent perspectives and strategies.

Those trained in the practice of design might be recognized as experts in the process, while other scholars and specialists deepen the team's breadth and depth of understanding as we prospect from different areas of knowledge to identify possible futures and routes.

Distinct practical skills are also key. As we have witnessed at MIT, a design team might include roboticists and geneticists, astrophysicists and anthropologists. Stakeholders also play an integral role, although it is important to note that pushback from stakeholders needs to be carefully mediated. As experts in the local context, they participate from the very beginning of the process, when the vision is first developed and defined.

Such inclusion should not cause the design team to be inordinately large. A core of 4 to 8 team members is typically considered ideal, although the broader collective of participants in the process may extend further.

"The real value of design education isn't in specific knowledge — it's in the interpretive lenses you're given to navigate an uncertain world. Designers are given tools not to control the world, but to understand and shape it."

Stefano Mirti
Architect, educator, strategic design
Interview **42**

The Design Conductor

For the design team to operate as a functioning collective, as opposed to a segmented assembly of different perspectives, requires the involvement of a design "conductor." Use of the term in this context was coined by one of the coauthors of this book, Sigurdur Thorsteinsson, in collaboration with psychologist Gianandrea Giacoma. By providing a new way to think about so-called "soft skills" in relation to managing, energizing, and synthesizing teams and projects with many moving parts, the notion of a conductor is particularly useful when addressing supercomplexity.

As Giacoma explains, "A true conductor and complexity-ready team will evolve their own way of being, encompassing not only theoretical knowledge and techniques but ways to express them with implicit and explicit meaning. This is an area where psychology, education, complexity theory, and *progettare* converge."

The significance of the word "conductor" can be traced to its Latinate origins: *ducere* (to lead, to guide), *con-* (with). In other words, there is no conductor without conjunctions — without a network of connected and independent elements.

This will be a truism for anyone who has attended a concert and studied the movements of the conductor in relation to the orchestra. When all parts are moving fluidly together, it becomes impossible to distinguish between who leads and who is led. A gradual crescendo on the part of the violins prompts the conductor to sign for a rise in tempo; this, in turn, signals the entry of the bassoons, and the conductor mitigates their impact with a gesture indicating less vibrato. In this way, the performance continues in a constant flow of interdependent communication (see Figure 5.1).

Likewise, a skillful design team is inclusive and diverse by nature. We do not use these two words lightly. In our commitment to look "under the hood" of the design process, to investigate, clarify, and apply the "why" and the "how" of design, we have come to recognize that the proof is in the practice. Our interpretation of inclusivity and diversity are not symptoms of the "morality of the moment," but a pragmatic recognition of how consensus solutions are reached.

Leadership in Flow

"Designers are not a mystical breed of beings that emerge, but they are made in the right circumstances. Designers engage with people and facilitate the discovery of new solutions that are waiting to be uncovered."

John Ochsendorf
Engineer, designer, educator
Interview **48**

Through our own practice, and in conversation with designers and innovators, we have observed that polymathic teams are most effectively led by a person or persons fluent in design methods and processes. We believe that such experienced individuals are best positioned to inspire enthusiasm and trust in the process, anticipate problems, and seize opportunities.

That said, experience and expertise need not automatically convert to hierarchical organization. Instead, by taking on the position of a conductor, such leaders can work strategically in flow, serving as a source of centrifugal energy and direction for the team as a whole.

INTERVIEW FINDINGS

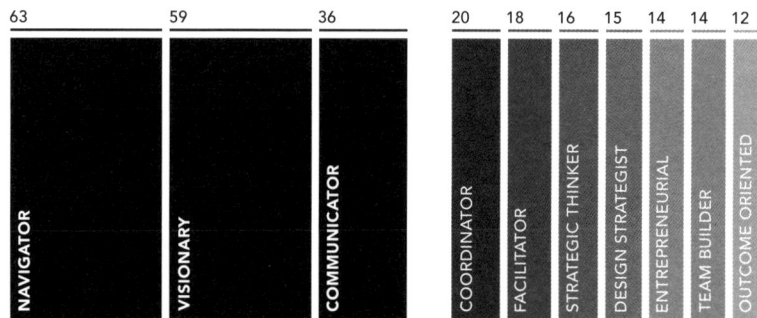

| 63 | 59 | 36 | | 20 | 18 | 16 | 15 | 14 | 14 | 12 |

NAVIGATOR / VISIONARY / COMMUNICATOR / COORDINATOR / FACILITATOR / STRATEGIC THINKER / DESIGN STRATEGIST / ENTREPRENEURIAL / TEAM BUILDER / OUTCOME ORIENTED

Figure 5.1 | Design Conductor Skills. Data analysis of interview transcripts revealed three critical skills for design conductors. They must be skilled navigators, visionaries, and communicators to manage the design process. Numbers in the graph reflect ranking of concepts based on their frequency and contextual significance across interviews for this book (see Appendix, Methods).

If the leadership role of the design conductor is not hierarchically ordained based on expertise and experience, what validates their license to lead? The key is to cultivate and sustain a culture of trust, itself a process in flux. Continuing the analogy introduced above, we explore how mutual trust is developed in practice.

Comparable to the position of an orchestra conductor, the design conductor has a clear sense of what the whole should achieve, as well as each moving part of the project (or performance), while simultaneously understanding the potential of each player and their instrument. Ambidextrous and emotionally intelligent, the conductor harmonizes diverse perspectives, disciplines, and objectives toward a unique interpretation of the music (as played rather than written).

Unlike conventional project management, the role of the design conductor transcends mere coordination; it entails synthesizing insights from various domains, understanding human behavior intricately, and bringing emotive resonance to plans and sketches (comparable to sheet music). The design conductor fosters a polymathic approach to decision-making and coordination by promoting mutual respect among team members with different backgrounds and skills, leading to richer tonalities and a unified outcome.

"Leadership comes intuitively through years of experience, which gives designers the courage to travel down the unbeaten path. It's like they activate everyone in a room through their complete trust and love of the design process, which has never let them down. At the same time, they also make space for that one aberration where it's the youngest person in the room or the least experienced person who has the most brilliant idea. The idea that's going to change the game."

Peter Coughlan
Strategic designer and systems innovator
Interview **20**

"For me, design is choreography because choreographing typically involves humans and bodies. The minute you apply the word design to people it is going to require a lot of precaution because different people have different perspectives, different values, different needs, and different personalities; different in every way, so you have to harmonize."

Nathalie van Bockstaele
Researcher and choreographer
of speech and action
Interview **62**

The orchestra conductor spends months tirelessly rehearsing, challenging the players to achieve their highest level of performance. Incorporating inputs from all involved, the conductor choreographs a path to realizing the vision of the whole, employing various methods to constrain and focus the attention of all upon that end. Indeed, "choreography" can be understood as one function of the conductor, physically engaging with the flow of the music through movement and gesture, infusing it with meaning in a way that can be emotionally read by the orchestra.

Likewise, the physicality of design is central to the process. Effective design conductors are typically highly active. Far from solely an intellectual activity at the desk, you'll find them constantly moving through the workspace, diagramming, illustrating, analyzing data, and synthesizing concepts with the group.

In this flow of interaction, the design conductor is more than a neutral "mediator." They are an active participant with their own distinct point of view, passion, and stake in the process, often a key player inspiring the original project vision.

The Optimism of Making

By iteratively refining their understanding and approach, the design team and design conductor co-evolve with the problem they are seeking to address. The dynamic interplay between vision, context, and potential can build enthusiasm for the process, transforming individual perceptions into a shared outlook and making the journey toward the solution as rewarding as the destination.

The task of the design conductor changes as the team travels down the vortex through cycles of synthesis (as outlined in Chapter 4). In the Discovery cycle, the focus is on instilling confidence that change is possible as we envision a more desirable future, and fostering a mindset of curiosity when prospecting the problem–solution space for relevant facts, ideas, and opportunities.

As the design team moves into the Framing cycle, the design conductor's task shifts to organizing disparate nuggets of information and insight into coherent clusters that can be storified. Finally, in the Emergence cycle, the role of the design conductor shifts to advocating discipline and decisiveness in selecting the most promising storylines to follow and evaluating choices encountered along the path.

To cultivate a team responsive to these tasks and ready to move forward into unknown territory, the design conductor must pay special attention to supporting the conditions for a culture of creative optimism.

This is an environment infused with faith that the team will eventually succeed; a workplace in which every team member feels respected and heard. In which ideas keep flowing, feedback is valued, and inevitable failure is framed as a learning opportunity, not a setback. However, bearing in mind Harvard professor Gary Pisano's argument that "tolerance for failure requires an intolerance for incompetence; and collaboration and the willingness to experiment must be countered with individual accountability, rigorous discipline, and strong leadership" (Pisano, 2019).

All in all, conducting involves constant calibration; balancing direction with flexibility, providing emotional support and instruction through active engagement with the medium, and empowering individuals to contribute their own unique skills, experience, and interpretation to the vision. As many have experienced, there is joy to be found in making things together.

This amounts to using all the power of synthesis that the collective and their conductor are capable of — not only reading the musical notes or design drawings (relying on the left hemisphere of the brain) but feeling the meaning and emotion behind what is signified (the domain of the right hemisphere). This way of working encompasses intelligence and emotive resonance, connecting the team through a shared sense of purpose and mobilizing action toward an ambitious whole.

Key Attitudes

In previous chapters, we have outlined the primary differentiators of the design process and why it is ideally suited to engage supercomplexity.

Design takes advantage of non-systematic thinking to solve problems and discover relationships that can only be understood from a human perspective. It enables practitioners to recognize patterns and move beyond data, using iteration and visual language to find common factors that resonate with all team members. In the context of supercomplex problems, design does not strive for a single optimal solution but can serve as an instigator and facilitator of dialogue and polymathic collaboration. These processes weave together in the pursuit of a preferred solution — that which is best aligned with the originating vision.

In our research, the following key attitudes stood out as foundational to the practice of design when engaging supercomplexity: the capacity to be daring, intuitive, and reflective.

Designers must dare to go beyond present limitations, trust intuition to innovate ways forward, and act reflectively to ensure effective and ethical outcomes. They have the courage and optimism to take calculated risks; the openness to an intuitive understanding about which direction to take; the breadth of perception to carefully examine the potential effects of a choice upon all parties involved.

These implicit, intangible qualities are often described as "soft skills" — interpersonal and behavioral attributes that enable effective collaboration. Unfortunately, this terminology has been problematic in the context of our left-brain-oriented culture, which favors conscious instrumentality over tacit knowledge and intuitive wisdom. Nevertheless, the soft skills embedded in these highlighted attitudes are critical to the design process.

"You have to allow space
for the risk factor, for
the unconventional, for
the unknown."

Sigrún Birgisdóttir
Architect, educator
comparative cultural studies
Interview **10**

ATTITUDE 1 | DARING

The quality of being daring is of particular importance to this book. It can manifest in various aspects of life: launching entrepreneurial ventures, pursuing artistic endeavors, advocating for social change, or embracing personal transformations.

To be daring involves the courage and optimism to take calculated risks and pursue paths that may be uncomfortable. It is a testament not only to curiosity and ambition but also to a sense of adventure.

Designing for the unknown and unexpected requires the grit to envision radical solutions, challenge conventional thinking, explore unfashionable ideas, and take risks in pursuit of positive outcomes. This may expose the designer to criticism, cynicism, and the judgment of failure from the perspective of other disciplines or, more painfully, from the design community itself. Moreover, daring requires the nerve to raise potentially obvious and obscure questions that others have overlooked and to have the humility to put aside personal preconceptions and follow wherever the answers lead.

In her lifelong work on leadership, social psychologist Brené Brown defines daring as a willingness to be vulnerable:

The courage to be vulnerable is not about winning or losing, it's about the courage to show up when you can't predict or control the outcome… "People are opting out of vital conversations … because they fear looking wrong, saying something wrong, or being wrong. Choosing our own comfort over hard conversations is the epitome of privilege, and it corrodes trust and moves us away from meaningful and lasting change (Brown, 2018).

What motivates the capacity to dare, despite the fear of obstacles or potential failure? Those who are daring exhibit a willingness and capacity to grow, evolve, and create new knowledge when existing data or methods are insufficient. Their endurance may be founded in a desire for positive change, a conviction of the importance of engaging the problem, trust in the process to produce desirable outcomes, or simply a love of adventure for its own sake.

Ultimately, daring behavior can inspire others to courage, initiative, and willingness to engage challenges that conventional reason would avoid. They are drawn to explore uncharted territories, whether in personal, professional, or creative pursuits, driven by a desire for growth, innovation, or making a positive impact.

ATTITUDE 2 | INTUITIVE

Intuition underlies all genuinely new discoveries. While the capacity to be daring involves striking out into unknown territory, intuition can be understood as the initial clue or sense of rightness that provides the prompt to move forward. Whereas immediate instincts might not always be correct, intuition arises from a slow process of incubation as observations and ideas combine unconsciously in the mind — ready to emerge as an insight at the opportune moment.

For Isaac Newton, that inspiration took the form of a falling apple; Barbara McClintock's discovery of "jumping genes" was prompted by noticing the color patterns of maize kernels.; For Pablo Picasso, inspiration would arrive like a carrier pigeon. "The important thing is knowing that the pigeon has arrived," he reportedly told a friend. "You don't have to unroll the message and read it."

Picasso's remark implies the degree of trust that characterizes a receptivity to intuition; not the rigid conviction of the singular-minded, but an attitude of openness that recognizes the richness of what it does not yet know. Openness and curiosity also serve to refine and sharpen an intuition as it emerges.

The "eureka" moments idealized in the history of science are often presented as sudden and ready-formed revelations. They are more likely the product of sustained observation, conversations, past experiences, and the innate human ability for pattern recognition. When asked to explain intuition, we discover that we cannot rely on the

logic of explicit reasoning or consciousness. It turns out that one of our greatest resources at the core of human invention is not ours to control — it is a form of implicit knowledge that profoundly influences human thought, emotions, and actions.

The elusive and ever-changing nature of supercomplex problems means that prior solutions or established methods are rarely useful. We must therefore look beyond the sum of what can be known and measured. Likewise, as demonstrated in previous chapters, the practice of analysis is insufficient for engaging the scope of the different factors involved. Problem solvers are often faced with either a lack of data or a deluge of information from multiple fields of knowledge. Under these conditions, trusting one's intuition may be the only way forward.

Design takes advantage of intuitive thinking to solve problems, forge relationships, and negotiate shared outcomes in a fluid context; a process that requires continuous iteration and testing, as well as a synoptic view of all actors and elements. Intuition taps into our mental capacity to make connections and form dynamic and evolving patterns from disparate information — whether data sourced from a present situation or imagined information based on a vision of the future.

In this way, intuition allows us to anticipate issues, see possibilities, and form holistic and empathetic connections with users and stakeholders. Learning to trust and act on intuition is one of the greatest benefits of design education and can be honed with practice over time.

"Flowing with complexity takes a different kind of nerve."

Kevin Bethune
Author, design innovator
Interview **8**

ATTITUDE 3 | REFLECTIVE

"When facing a complex problem, it's easy to get lost in the details. Reflecting on the broader mission keeps you grounded."

Bill Patrowicz
Strategic advisor and innovator
Interview **51**

Design is at work throughout our everyday lives. It influences and is influenced by society, acting to shape cultural norms, processes, and social understandings. Reflection on the meaning and fit of design practice according to a specific context is therefore crucial to successfully engaging supercomplexity.

The term "reflective practice" was coined by the philosopher and city planner Donald Schön in his book *The Reflective Practitioner* (Schön, 1984). It refers to a continuous process of self-examination and learning from experience:

Reflection-in-action occurs while a course of action is being implemented during the evolving design process, allowing individuals to adapt and adjust their approach in real time based on feedback and insights.

Reflection-on-action involves retrospective analysis after an action has taken place, providing an opportunity to dissect past experiences, identify areas for improvement, and extract valuable lessons for future endeavors.

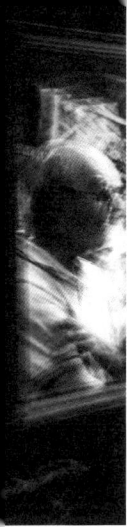

> "To be subversive with good intent is to go beyond empathy for the end user — it means understanding how incentives, politics, and internal dynamics shape the whole system."
>
> Kevin Bethune
> Author, design innovator
> Interview **8**

By embracing both forms of reflection, individuals cultivate a deeper understanding of their own role in enabling change and its subsequent impact. Reflection serves as a self-correcting mechanism of quality control, encouraging ethical behavior through the conscious self-monitoring of the consequences — both before and after a choice is made (including the choice not to take an action). The aim is to evolve solutions that not only solve problems but also promote social, cultural, economic, and environmental equilibrium. In this way, design becomes a quiet force for good, weaving together innovation and integrity to leave a legacy that uplifts people and the planet.

Subsequent methodologies based on Schön's research have developed (Brookfield, 2009; Gobbo, 2011). They all rely on an important differentiating feature of design; the premise that opinions, values, and feelings must be given equal credence to measurable facts. The participation of stakeholders from the beginning of the design process provides a crucial source of information that helps to uncover the holistic context of a problem.

> "There's a partnership dimension to good design... we're all in this, co-creating, coming from our place of expertise."
>
> Lisbeth Shepherd
> Social innovator
> Interview **56**

141

Without this foregrounding of stakeholder involvement and the validation of intangible elements, it is difficult — even impossible — to propose solutions with sufficient currency to be adopted and implemented. Any proposed path must take into account the well-being and rights of individuals, communities, and the broader society in decision-making and actions.

Specific to each context, reflection applies principles, values, or behaviors that are judged to be ethically aligned with locally accepted standards of conduct rather than measured against generalized abstractions of right and wrong.

It is important to note that reflection goes beyond the practice of "empathy" associated with Design Thinking, a popular management tool, which foregrounds the designer's eye in a process of close observation seeking to understand the needs of an "end user." In contrast, the practice of reflection recognizes a broader range of stakeholders beyond those who are most directly affected. Furthermore, empathy implies that stakeholders are outside the process, to be studied and appreciated by problem solvers who have the ultimate power to make decisions about the design. For all the claims of "beginner's mind" (open to many possibilities) often evoked by Design Thinkers, this attitude has been criticized for contributing to an elitist view that only trained professionals can solve supercomplex problems (Iskander, 2018).

As we have argued in the sections on the design team and the design conductor, expertise and experience in the design process are important. However, those skills can only be leveraged with a receptivity to the many resources of information and ideas within the problem–solution space. Reflection affords designers the space to step back and assess their decisions based on the specificity of the context, rather than by the ledger of their own expertise.

These attitudes play an active role in shaping the world around us, flexing the confines of individual and collective perception, and therefore the scope of possible solutions available. They reframe the way we interpret challenges, guiding decisions and influencing interactions with others. Design affords its practitioners this freedom to change and to grow with the same fluidity of a process that "creates as it goes."

INTERVIEW FINDINGS

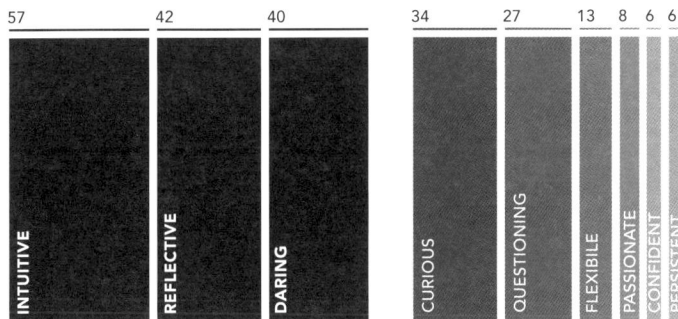

| 57 | 42 | 40 | 34 | 27 | 13 | 8 | 6 | 6 |

INTUITIVE REFLECTIVE DARING CURIOUS QUESTIONING FLEXIBILE PASSIONATE CONFIDENT PERSISTENT

Figure 5.2 | Attitudes. Data analysis of interview transcripts revealed Intuitive, Reflective, and Daring as key attitudes to engage supercomplexity. Numbers in the graph reflect ranking of concepts based on their frequency and contextual significance across interviews for this book (see Appendix, Methods).

"Integrative intelligence — like being able to create connections from different fields, different collaborations, different scales, different domains… enables new futures to emerge"

Skylar Tibbits
Designer, computer scientist
Interview **58**

For as long as problem-solvers can resist relying upon preconceptions (applying the enabling strategies outlined in this book), it is the nature of perception to expand and change in sync with new contexts and information. Transforming ideas into possible new futures and solutions requires innovation within the process itself.

Designers simultaneously initiate, shape, and experience the process of moving from the present to an envisioned future state. Everything — from the attitudes and techniques applied, to the formation and implementation of the eventual preferred solution(s) — is emergent.

It is this premise that redefines the role of the designer. No longer an individual or an all-seeing eye, but an interdependent factor of flow in a process that surpasses human intentionality and control. As such, the role of the designer is commensurate with the world in which we live. One in which emergent futures are both created by — and exceed — what we can perceive.

Beautiful Solutions

"Understanding the
limits of design is as
or more important
than understanding
what we can do."

Nicholas de Monchaux
Designer, author, educator
Interview **24**

Ultimately, the practices and attitudes explored in this chapter make way for solutions that could be described as beautiful.

"Beauty" is a contentious term in the context of design. While mathematicians and scientists use this term freely to describe the elegance and simplicity of an equation, not many designers dare to assert the value of beauty in their work; it is considered more credible to focus on utility or innovation. As alluded to in Chapter 2, the doctrine of "form follows function" advocated by 20th-century modernism has discredited beauty in design, rejecting it on grounds of subjective preference and relativism, more decorative than substantive.

However, what if the so-called "eye of the beholder" was understood not as a personal judgment of preferences and biases, likes and dislikes, but the perception of a process? The apparent beauty of a solution would therefore be judged by the way a positive outcome develops from multiple levels of synthesis, no longer a static entity but an evolving system with its own internal momentum.

In this way, an event — the process of making — might be described as beautiful; a quiet force for good, in which form follows meaning as much as function. A process rooted in responsibility to people and planet, rejecting singular heroic solutions in favor of sustainable systems that outlast their makers.

And so, the association of design with aesthetics comes full circle. No longer an optional add-on of imposed allure, but the source, orientation, and test of the X.

Learning Beautiful teaches computer science fundamentals to children through hands-on, screen-free toys made with natural materials.
MITdesignX 2017

CASE | **EDUCATION**

Learning Beautiful
Preparing for a Digital Future

To prepare children for a digital future, it is critical to start at a young age when foundational learning occurs. In the United States, 53% of public high schools offer computer science education of some kind (Code.org, CSTA & ECEP Alliance, 2022), yet primary schools struggle to provide curriculum and resources for K–8 classrooms. Since 2016, with the Obama administration's Computer Science for All initiative, there has been a nationwide effort to increase that capacity by extending the computer science education pipeline to lower grades. However, existing approaches use digital tools and screens, which evidence has shown to be harmful to young learners. Physicians warn parents of the negative effects of screens, and the American Academy of Pediatrics recommends limiting exposure (Griffin & Williams, 2024).

Learning Beautiful, founded by media artist Kimberly Smith Claudel, offers a solution to this paradox by providing children aged 3–9 with non-digital materials and curricula to learn the fundamentals of computational thinking. Developed in collaboration with computer scientists and experts in early childhood education, tactile learning materials resemble traditional child-friendly toys made of naturally finished wood and beautiful colors. The goal is to break down complex, abstract concepts of computation into simple building blocks appropriate to different learning styles. The innovation builds upon the insight that physical interaction is crucial to learning; children grow to understand spatial and social relationships and how to process inputs from the world around them.

Learning Beautiful offers comprehensive teacher training for educators to incorporate the materials into standard curriculum and classroom settings. First introduced in libraries, the materials are now being adopted by school systems and parents nationwide.

145

6   The Path Forward

Design serves to condition, predict, and produce new realities. No longer limited to the constraints of present thinking, we are free to invent a way forward

The journey of *Designing the X* has taken us through the cycles of the vortex. Now, by combining multiple perspectives and testing different ideas, we find ourselves following an emergent path forward.

As we have argued throughout the book, design can be understood as a powerful form of reasoning based on processes of synthesis rather than a reliance on analysis alone.

Instead of proposing that any single discipline should be prioritized, we envision a future of polymathic practice in which analysis is complemented and extended by synthesis, integrating knowledge derived from measurable data as well as less tangible human factors.

Design is not the sole "answer" to engaging supercomplexity and tackling the vast array of interwoven problems that threaten the sustainability of life on Earth. Rather, it is the key to unlocking the potential of the innate capabilities and tools we have evolved as a species, investing our technology with humanity and infusing meaning and intention into the objects, environments, and systems we create.

Crucially, design is also a means of taking stock of a contemporary problem and projecting alternative future states in which the problem does not exist. By giving equal credence to these visions as potential new realities in relation to the provable facts of the present, designers are empowered to work both pragmatically and optimistically, progressing through the various cycles of synthesis (outlined in Chapter 4) to find a pathway between the present moment and a better future.

Within a network of multiple disciplines, design serves to condition, predict, and produce new possibilities for our civilization. No longer limited to the constraints of present action and thinking, we are free to invent and act otherwise.

Redefining the Real

"The true quality of design lies not in the object itself, but in its ability to rewire the systems it enters and reframe the culture it touches."

Ezio Manzini
Designer, author, and
social innovator
Interview **40**

The stakes are increasing with the rate of advancing supercomplexity.

We can no longer afford to focus solely on the limiting certainties generated by so-called "left brain" analytic thinking, at the expense of "right brain" synthetic thinking, which allows for invention, compassion, and openness to the unknown.

Take the conditions of global warming, for instance. Science can measure and predict these phenomena with great precision; however, the data have failed to make the necessary impact on collective behavior or convince individuals and communities to adjust daily habits for the benefit of future generations. Rather than working forward to synthesize an alternative future, science typically focuses on explaining the phenomena of the present.

Furthermore, science is not fully equipped to effectively address supercomplex problems, given that it typically excludes factors such as human motivation, prejudice, and emotion. It is risky to dismiss these less measurable and controllable aspects of lived experience, not least because they demonstrate the effects of human perception in shaping apparent reality.

This can be witnessed at multiple levels, from the psychological to the neurological, and even at the quantum scale. The world we perceive is partly of our own design.

New discoveries are revealing that our sensory experience does not equate to the physical properties of the matter around us. For example, recent theories suggest that what we witness as "color" is an interpretation prompted by the brain as it acts to resolve differences in brightness; the polychromatic world is an illusion of the sense-seeking mind, constantly working to fill gaps and inconsistencies for the sake of our survival (Moddel, 2023). Likewise, what we experience as thoughts, sentences, or music emanates from the organic structure of the brain. "The eye sees the stars and the brain sees the constellations," explains linguist and neuroscientist Andrea Moro, referring to his recent book coauthored with Noam Chomsky (Chomsky & Moro, 2022).

At an even more fundamental level, it is worth considering the revolutionary challenge to reality that is taking place in the field of theoretical physics.

Beginning in the late 19th/early 20th century with the advent of quantum mechanics, it was revealed that matter at the atomic and subatomic scale appears to defy accepted physical laws (O'Connor & Robertson,1996). This is a realm where quanta of matter can evidence themselves as particles or waves, depending on how — or if — they are observed, while these same particles instantaneously react with other "entangled" particles vastly distant in space or time (Cramer, 2019).

"Multiscalar thinking calls for cultivating dialogue between the parts and the whole — recognizing, through continuous feedback and co-evolution, that the whole shapes the parts just as the parts shape the whole."

Luisa Damiano
Epistemologist of complex systems philosophy of emerging sciences and technologies
Interview **23**

Furthermore, the theory of "superposition" — which indicates that matter can take multiple forms and positions simultaneously, only resolving (or "collapsing") into a measurable state at the instant of observation — has become a cornerstone of contemporary physics, thereby positioning human perception at the center of what is considered to be real (von Neumann, 1932). Prior to the act of observation (according to quantum theory), reality exists only as a set of probabilities concerning when, where, and how it may be manifested.

This is a radical departure from the Newtonian framing of the universe as a set of fixed forces — indeed from any ideal of an orderly cosmos of discoverable laws external to human existence (Harrison, 1986). Even Einstein resisted this new framing, exclaiming, "God does not play dice with the universe" (Born, 1971). Nevertheless, quantum theory has proven remarkably accurate in predicting the composition and behavior of matter at the nano-scale (Albert, 1992).

You might expect such a revelation to entirely change the way we go about our lives, suddenly aware that reality is shaped by perception and essentially "conditional" (as the philosopher Immanuel Kant argued in the 18th century; Stang, 2024). Change is shown to be the only constant; the ground under our feet has become a process of flow.

Redefining Reason

"Individuals who think and act
with artistic sensitivity are
not just contributors — they
are catalysts. Their ability to
imagine alternative futures,
challenge assumptions,
and provoke meaningful
questions brings depth and
direction to projects to go
beyond solving problems."

Shunji Yamanaka
Design engineer, mechanical and
biofunctional systems
Interview **66**

As we have demonstrated in previous chapters, the design process is uniquely equipped to work with this flow.

When a problem presents itself, an initiating vision serves to open a conceptual space — a window of opportunity in the commonly accepted flow of reality — which allows the design team to prospect and frame clues and ideas toward a solution.

This sets in motion a series of events (the vortex) in which the vision can evolve to become actionable. A new reality can emerge through a process of discovery and invention that values human intuition, fictions, and alternate future realities, while recognizing those aspects of the vision that already exist in the present and will likely continue into the future. The process encompasses multiple perspectives and employs methodologies that increasingly constrain choices until a way forward emerges.

It is not too bold to claim that design is what we do when we are at our most "human." That is, at our most humane, reflective, inventive, communicative, and transformative. When we are acting with vision, persistence, and optimism in relation to an open and ever-altering interpretation of reality.

To quote a phrase attributed to Aristotle, our species can be differentiated as "rational animals" (Keil & Kreft, 2019). Beings with the capacity to reflect on our condition and take intentional action that serves to change our environment — and ourselves. This is how we "co-evolve" with nature. Rather than being passively swept up in the process of evolution, we play an active role by creating the tools, both hardware and software, that adapt us into new versions of "being human."

The trouble we face — the unstuck myth — is the impoverished interpretation of rationality that has served to limit human possibility, causing a ripple effect of destructive consequences. "Reason" is not limited to the analytic segmentation and recombination of tangible facts and provable axioms; the practice that has come to be associated with the dominant fields of science and technology that currently drive the global economy and accepted forms of problem-solving.

If the capacity to "reason" differentiates us as *Homo sapiens*, then wouldn't it simply amount to applying the full spectrum of capabilities in our human toolkit? It is now increasingly clear from insights gained by the natural and social sciences, as well as the arts, that our capacity to make meaning and take action is far richer than what dominant fields of practice would have us imagine.

The Next Renaissance

A precedent for attaining a more expansive vision might be found during the period of the Renaissance in Europe, before the establishment of the fully-fledged scientific Enlightenment of the 17th and 18th centuries. Originating in the 14th century and sparked by a revived interest in the classical world, the Renaissance was characterized by a polymathic curiosity for the full breadth of human experience, resulting in innovations such as Brunelleschi's discovery of linear perspective and Copernicus' radical reframing of a heliocentric universe.

Of course, neither discovery is the full story, and the scientific Enlightenment did much to debunk faulty assumptions and impasses of perception. However, while the word "enlightenment" hubristically proposes a point of arrival, the word "renaissance" ("rebirth") implies a continual process of becoming. The latter phrase is better aligned with the inherent dynamism of our evolving planet. Today, we are entering yet another renaissance.

Liberated from its "upper case" status and in flow as one among many futures, what might our current renaissance emerge into? In the European Renaissance spanning the 14th and 17th centuries, the guiding motivation was to release intellectual inquiry and definitions of reality from the dogma of the church and crown. Even as wars of religion persist and new versions of feudal hierarchy continue to sculpt social life, human citizens today are comparatively free to ask questions, take action, and work collectively for positive change.

"Design, intended as reflective practice, can provide leaders a powerful frame to change the world."

Roberto Verganti
Scholar and author
Art and innovation,
design theory
Interview **63**

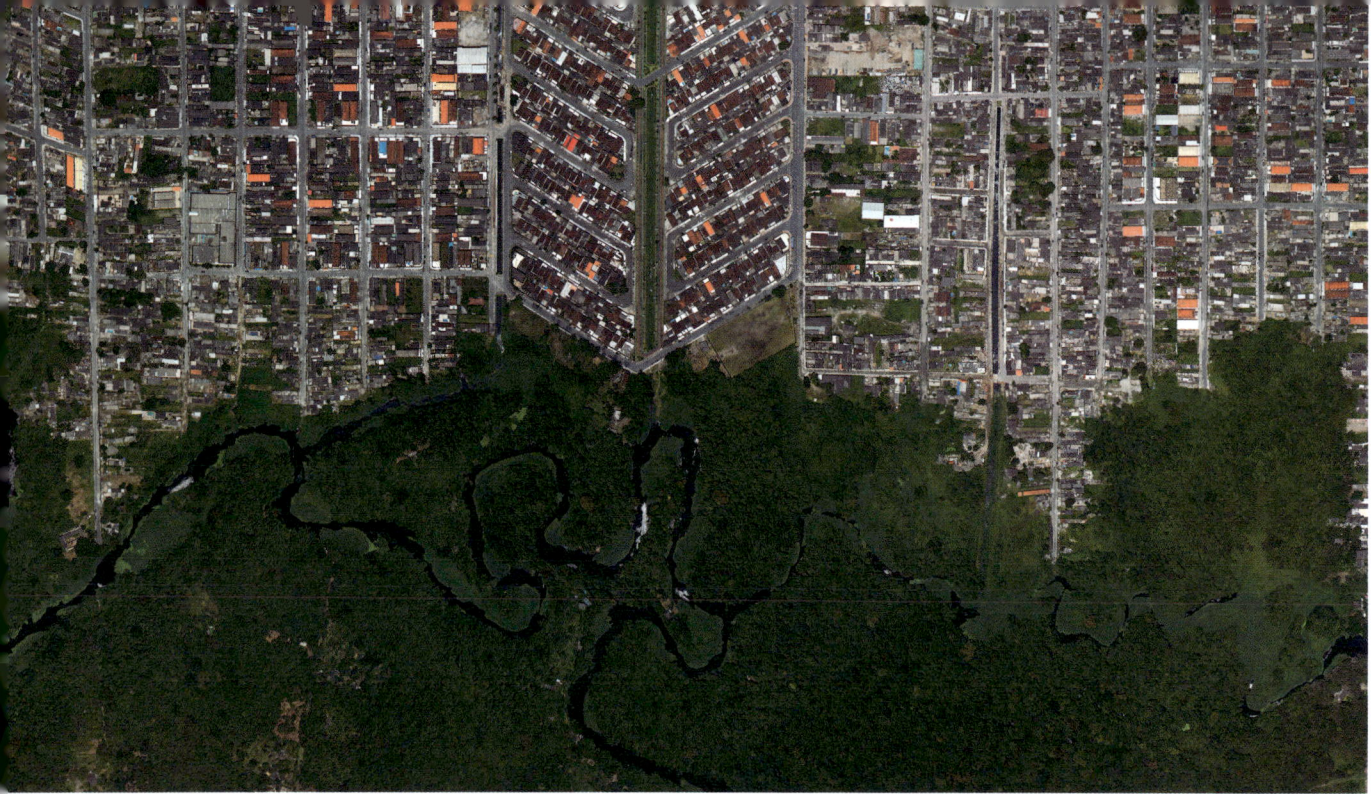

Vila Caiçara, Brazil, consuming the rainforest.

When a particular field of practice holds the purse strings of the economy, the values of a society have the potential to be shaped accordingly. However, even as the disciplines of science and technology condition the tools we use, the products we consume, and what is perceived as "truth," they are not "principalities and powers" to be defended against. Scientific discoveries and technological advancements have led to tremendous improvements in human living conditions, demonstrating exceptional feats of ingenuity, curiosity, and the drive to discover and create.

That said, we need only to look around us and witness the ongoing destruction of our Earth-bound habitat to recognize the methods employed by science and technology are certainly impactful — but not necessarily aligned with the delicate equilibrium that sustains life on this planet. Dwindling resources, global pollution, extreme urbanization, human conflict, mass extinctions, and increased risk of pandemics threaten to wipe out our species entirely. It is clear that a piece is missing in the puzzle of how to engage supercomplexity. As expressed by Terrance Deacon, "when unrealized future potentials appear to be organizers of antecedent processes that tend to bring them into existence, it forces us to look more deeply into the ways that we conceive of causality and worry that we might be missing something important." (Deacon, 2012).

Change Starts With Education

As educators and design practitioners advocating for the equal role of design in relation to the dominant STEM fields (science, technology, engineering, mathematics), we believe that design holds the key to unlocking valuable aspects of current practice. By drawing upon the insights and methods of multiple disciplines, as well as its own unique contributions, design acts as a catalytic agent to invent new ways forward,

This is distinct from attempting to make design more scientific, or science more akin to design. Our repositioning of design does not seek to replace or dominate other disciplines, but instead serves to synthesize them as part of an interrelated whole; an exchange of heterogeneous information and diverse modes of interpretation resolved by a process of evolutionary selection.

In our work as educators at some of the world's foremost institutes of design, we have had the chance to cultivate this vision of a polymathic environment of learning and engaging problems — designing a microcosm or prototype that can equally be applied to contexts of industry and public policy.

"What if the most unexpected images are the ones that provoke dramatic policy change? The power of imagery to shape society is often hidden in the most surprising places."

Caroline Jones
Art historian, curator
Interview **32**

However, it is also apparent that conventional systems of education (and the bureaucracies that uphold them) are thwarted by a bias toward stability and fixity. This is understandable, if not acceptable, given that all humans are inclined to seek simplicity and security in defense against the vertiginous stakes of supercomplexity.

From birth, children are conditioned to analyze their world by partitioning it into segments. Eons of time are compartmentalized into hours and minutes, the superflux of space is converted into feet and inches, and the world is measured against the manageable dimensions of the human body. People, likewise, become units of greater or lesser abstraction — the immediate family or caretakers holding more meaning and allegiance to the child than the "others" who pass by on the street. Over time, this leads to a "fallacy of completeness": a false perception that the artificial measures implemented by the "left brain" are the primary reality (McGilchrist, 2011).

When further instructed in analysis throughout the early years of education, the quantified "apples and oranges" and the "Ryans and Sams" of practice algebra problems convert all too easily to lived experience. People and produce alike are abstracted as data to manipulate, an attitude now rewarded by some of the highest-paying jobs at the foremost technology companies.

Assessing Competitiveness

Given that the merit of higher education is now quantified according to grades and student recruitment based on the promise of a ticket to a lucrative career, it is no wonder that human-centered learning (arts, psychology, literature, history) is in decline in U.S. schools and campuses (Kemsley, 2019).

STEM is considered to be a more financially viable track toward employment in a technology-driven world, with a "pipeline" of STEM students produced as ballast to defend U.S. dominance in science and technology. This attitude has its roots as far back as the 1940s and 1950s, when the anxieties of the Cold War prompted mass federal investments in STEM in support of the defense industry. The surge in investment recurred in the 2000s after the bursting of the dot-com bubble and led to the COMPETES Act of 2007, which sought to enhance U.S. competitiveness by investing in STEM education, R&D, and workforce development (Mandt et al., 2020).

Nevertheless, research has indicated that the effective value of STEM skills (as defined by financial gain and flexibility within the marketplace) diminishes quickly in the years following graduation due to rapid technological change. Furthermore, relative employability balances out in the later years of a career, when those with social, humanistic, and collaborative skills tend to pull ahead. These individuals are the preferred candidates for jobs in management, strategic planning, and innovation — often, ironically, in technology companies (Deming & Noray, 2020).

The immense value of STEM disciplines is clear. However, there are also concerns that the failure to take account of the social context of human thought and motivation is leading to unintended and potentially highly damaging consequences. As a result, the National Academies of Science, Engineering, and Medicine are now advocating for a return to an approach to education that balances science with the humanities (Mullen, 2019).

This counter-trend provides an opportunity to introduce design as part of the core curriculum at all levels, reassessing what it takes to cultivate effective solutions. Educators are currently at risk of training the next generation with a limited toolkit (Norman, 2011) — sending them off to face new challenges with blunt tools and outdated mindsets, ill-equipped to successfully navigate toward a more sustainable and equitable future. While STEM focuses on analytical thinking, design education fosters synthesis and human-centered approaches.

"The way I see it, these questions of ethics and computation and AI are all questions of design because design is where the rubber meets the road."

Molly Wright Steenson
Historian of Artificial Intelligence
Interview **57**

Integrating STEM and design education enhances individual and societal competitiveness by combining technical skills with tools and methods to better understand the context and envision opportunities for problem-solving and innovation (Gale et al., 2020; Hallstrom & Ankiewicz, 2023). A design-informed approach would prepare individuals to tackle complex challenges, bridging the gap between technical solutions and human needs. At a societal level, this integration drives sustainable development, promotes cross-disciplinary collaboration, and enables communities to actively change present conditions and thrive in a rapidly evolving world.

A rebalancing of ethics and strategy needs to occur at the level of pedagogy — the how and the why of what we teach. By applying methods of synthesis to progress, advance, and complement the knowledge gained by analysis, it is possible to complete a circle of reasoning that makes full use of the human capacities to investigate present conditions and to synthesize alternative, more desirable futures.

In science, it will be vital to recognize the intangible elements of human experience — taking account of the less measurable variables in our perceptual framework when formulating theories for describing (and thereby acting upon) the physical world (Deacon 2012). In design, the challenge will involve prospecting for enduring elements of the context likely to continue into the future. To increase the potential to actualize their visions, designers must work with those elements of continuity as well as activate possibilities for innovation. It is important to recognize that much of the future already exists in the present.

At this critical point in human development — when we risk being superseded by our technology — we still have the chance to turn and reassess current frameworks of perception and practice. In a world where analysis has become the dominant definition of human rationality, it is little wonder that anxieties are rising about the potential for human obsolescence due to the powerful analytic machines we have created.

Against Obsolescence

"Technology amplifies human capabilities, so if society has biases, technology will amplify those biases. We must stay vigilant and actively address these issues as they emerge. Technology is not predetermined — it is our responsibility to shape it intentionally and with care."

Marcelo Coelho
Computation artist and designer
Interview **19**

This book is not a comprehensive roadmap for how to achieve an alternative future. Rather, we propose a vision and seek to initiate a much stronger exchange between the two epistemologies of analysis and synthesis, following the processes of the vortex to evolve a path forward.

A starting point for that emergence can be found in the motto of MIT, "mens et manus." Often understood to simply mean the integration of theoretical investigation and hands-on learning — the difference between solving a problem on a blackboard and building a functioning machine — the implications are much more neurologically complex. It has been noted that the felt knowledge generated by working with the hands serves to stimulate speculative ideas latent in the mind, and vice versa. It is from this process of synthesis that the most successful advances emerge — those that actively change how we evolve as a species.

The question is whether those advances and that direction of change are in service to human and planetary flourishing. This is where the ethical role of design becomes clear, acting as a catalyst for synthesis in all disciplines. Design, we argue, is inherently humane; both on account of being an innate human capacity and due to its dependence upon context (the reality of situatedness and lived experience) to inform decisions. In design, detached abstraction is not an option.

A Program for Change

In this section, we present a program for sharpening the tools at our disposal to better equip the next generation to recognize and work with supercomplexity. Design trains practitioners in the following starter skill set:

Synthesis — inventing new wholes that are greater than the sum of their parts, rather than focusing on analysis of the parts alone.

Critical thinking — the daring and curiosity to question accepted norms.

Collective decision-making with care — the commitment to mutual empowerment of disciplinary experts and stakeholders for positive ends.

Intuition — recognized as of equal worth to measurable facts.

Persistence in flow — the capacity to work on a problem that has no single "right" answer, in the midst of changing circumstances.

Mastering such skills will require channeling education toward:

Project-based learning — tackling real-life complex problems in classwork and in the lab.

Teamwork — as opposed to individual achievement.

Making — as a companion to studying.

Polymathic learning — synthesizing diverse areas of knowledge.

Student-centered learning — empowering students to direct their own learning process.

A future-focused vision — emphasizing innovation and unconventional pathways for change, rather than parsing the past.

To benefit from these new competencies and to foster the further development of graduates, change will be equally necessary in industry. The alternative is for companies to continue to hire the next generation of problem solvers into obsolete organizational structures, conditions ill-equipped for increasingly complex situations.

As is often the case, the resistance to change can be traced back to education systems. Enormous amounts of money, brand value, and personal human capital have been invested in present-day curricula. This brings us to one of the principal motivations for this book: to widen the parameters of design education, expand existing models based on industrial production, and move toward a focus on engaging supercomplex problems.

In too many cases, the pedagogy of design teaching is still entrenched in the modernist tradition of the early 20th century (Norman, 2011). Design students are still to be found taking on excessively long hours, working alone on a hypothetical problem given by an instructor who acts as the sole evaluator of success. Both the question posed and the method for answering it are often divorced from social and economic circumstances and direct involvement with the context. We question whether this is the right strategy for training the next generation.

Based on the findings of our research and practical experience, we propose the following recommendations for preparing young designers for the world they will enter. In addition to educational settings, these principles are equally applicable in a professional environment of team leadership:

1. **Remove the role of the "master,"** and teach designers as partners in the process. This will promote mutual learning, alignment of perspectives, and development of the soft skills necessary for shared decision-making.

2. **Create the conditions for polymathic teamwork** in studio settings.

3. **Teach basic methodologies of synthesis**, and how and when to apply them in the vortex of the design process.

4. **Frame the learning process as the hands-on practice of a craft** as opposed to a cerebral exercise.

5. **Enable designers to define their own problem space** by first envisioning a preferred future state.

6. **Encourage designers to develop a deep understanding of the context** by prospecting for clues across multiple fields. This includes known facts of the present, as well as ideas, concepts, and opportunities for change in the future.

7. **Encourage entrepreneurial thinking and risk-taking** in facing the unknown. It is better to fail and start over than to continue with a safe approach that may be of little value in the future.

8. **Maintain a steady orientation toward the future**, undaunted by accelerating change. In a world where the extent of human knowledge is estimated to be doubling every 12 hours (Schilling, 2013), it is impossible to design for the present.

9. **Accept that problems may have more than one "right" answer.** Rather than optimization (in which goals are traded off against each other), the aim of design is to fulfill the holistic intent of a vision. This may be attained via different routes.

10. **Recognize that the studio is not the only zone of design education**. Whether from alternative subject areas or personal projects, designers must have the chance to follow their passions and draw upon knowledge from multiple fields.

11. **Provide space for reflection in the curriculum** (and/or design process), where designers can assess their own impact and the effects of their ideas on society and its context.

12. **Cultivate a mindset in which design is an act of care rather than control**, founded on the commitment to create positive change.

"The true value of design is rooted in its capacity for lasting impact. When design is crafted collectively with depth, intention, and a clear understanding of the people or challenges it seeks to address, it not only resonates more powerfully — it endures."

Andrea Chegut
Innovator, research scientist, and financial econometrician
Interview **16**

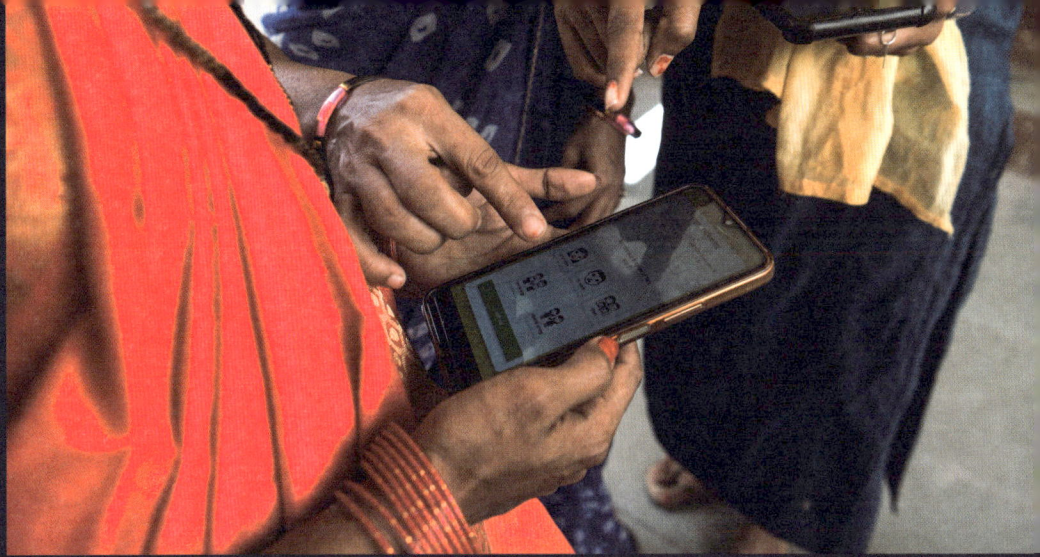

Bandhu provides access to jobs and safe, affordable housing for migrant workers in India. MITdesignX 2019

CASE | **MIGRATION**

Bandhu
Reforming Urban Migration

Easing the pain of human migration is one of the greatest global challenges; migration and its human toll will continue to increase dramatically in the years to come (McAuliffe & Oucho, 2024). Media narratives of refugees fleeing war and injustice are only part of the story. Seasonal migration for economic survival has become a way of life for populations worldwide.
In India, over 400 million internal migrants traverse long distances in the off-season to find interim work in cities so that their families at home may be supported (Kaushik & Campbell, 2023). These workers are vulnerable to exploitation by local agents who charge high fees for access to low-paying jobs and substandard living conditions — a place to sleep on the ground or in overcrowded rooms with no sanitation and little to no personal security. Founded by real estate developer and urbanist Rushil Palavajjhala and data scientist Jacob Kohn, Bandhu offers an AI-driven platform and locally based educational, social services, and advocacy for families to ease the challenges of migration

The company, supported by the Gates Foundation, received a Sheltertech Award for making the rural-to-urban migration process more equitable. Rather than traveling hundreds of miles, waiting days in the sun for a chance to secure poor housing and jobs, workers can make informed decisions from home about where, when, or even whether to migrate. Through feedback loops, unscrupulous employers and landlords are identified and removed from the platform. With a vision to reform rather than replace migration patterns, Bandhu has succeeded by partnering with local agents, migrant community groups, and NGOs on the ground at both ends of the migration journey, harnessing their experience, contacts, and data for social good. Bandhu now serves users with thousands of jobs and 250,000 beds offered on the platform. Already a key player in India's digital public infrastructure, the aim is to expand the platform to serve migrant communities worldwide.

Epilogue

With this book as a touchstone, a set of proposals to hand, and millennia of human invention as a precedent, we invite you to step into a world of supercomplexity. A world where you feel enlivened by the unceasing dynamism of your surroundings and an expanding range of possibilities.

As with any process of emergence, you will likely have been changed along the way. Arguably, this is why we pick up a good book, watch a film, embark on a journey, or pursue a process of making. Not to have our expectations confirmed or to secure our world in stasis, but to be changed — and to become more like ourselves in the process.

This is the paradox at the heart of so many pilgrimage narratives — from Dante's *Commedia* (Alighieri, ca.1320/2002), *The Epic of Gilgamesh* (ca. 2750–2500 BCE/2003), and *The Bhagavad Gita* (ca. 200 BCE/2007). We end where we began. The resolution of the vortex becomes the source. The Earth, our home and grounding, is recognized anew — a place of tremendous richness and complexity where we are all contingent agents of transformation, changing and changed by the environment in which we live.

The grounding of this return will only sharpen individual perception of the challenges to life on this planet. However, as this book has sought to prove, we are each highly capable of acting, adapting, and ultimately transforming our shared habitat. In this way, we evolve, consciously participating in the extraordinary phenomenon of synthesis and emergence.

The path forward is also the path to the center and the source of the vortex — a geometry as iterative and non-discrete as a Möbius strip or a Klein bottle. And so, we keep moving. We keep designing, we keep evolving with the flow of supercomplexity. Design creates as it goes.

Authors

Dennis Frenchman

Professor Emeritus
Massachusetts Institute of Technology

Dennis Frenchman is Emeritus Professor of Urban Design and Planning at MIT, and the founder and Chair of MIT DesignX, the Institute's program for accelerating innovation in design, cities, and the human environment. At MIT he has served as the Associate Dean of the School of Architecture and Planning, Director of the Center for Real Estate, Chair of the Masters in City Planning degree program, and Head of the City Design and Development group.

As an architect and city designer, Dennis has a distinguished record of practice focusing on the transformation of cities. Most recently, he served as a senior partner in Tekuma Frenchman Urban Design, leading projects in Asia, Africa, the Middle East, and Latin America.

He is an expert on the application of digital technology to city design and has designed large-scale, media-oriented cities and industrial clusters, including Seoul Digital Media City in Seoul, South Korea; the Digital Mile in Zaragoza, Spain; Media City: UK in Salford, England; Twofour54 in Abu Dhabi, United Arab Emirates; Ciudad Creativa Digital, Guadalajara, Mexico; and the Medellín Innovation District in Colombia. He also served as an external advisor on urban livability to the president of the World Bank.

Dennis is the author of articles and books on advanced urban design, including *Technological Imagination and the Historic City* (2008, Ligouri, with William J. Mitchell et al.). He has a particular interest in the redevelopment of industrial sites and has prepared plans for the renewal of textile mill towns, canals, rail corridors, steel mills, coal and oil fields, shipyards, and ports, including many of international cultural significance. Recently, he led an MIT research effort to develop new models for clean energy urbanization in China, sponsored by the Energy Foundation.

His work has been widely recognized, including awards from Progressive Architecture, the American Institute of Architects, and three citations from the American Planning Association for the most outstanding projects in the United States.

Svafa Grönfeldt

Professor of the Practice
Massachusetts Institute of Technology

Svafa Grönfeldt, PhD, is a Professor of the Practice at the MIT Morningside Academy for Design (MAD) and a founding member and faculty director of MITdesignX, a design innovation program founded at the MIT School of Architecture and Planning. Her work focuses on harnessing the power of design to tackle complex challenges and drive innovation success, with an emphasis on sustainability, impact, and the development of high-growth ventures.

An entrepreneur and venture designer, Svafa operates at the intersection of academia and industry, collaborating with polymathic teams worldwide to design and build award-winning programs, brands, systems, and organizational strategies to drive business performance and workplace well-being.

Before joining MIT, she was a member of a team of entrepreneurs that scaled two global life science companies from the ground up, where she served as deputy to the CEO and Chief Organizational Design Officer. Between these roles, she served as president of Reykjavík University, where she led major institutional initiatives and strong collaboration between academic and industry partners.

Svafa serves on the boards of three publicly traded companies listed on Nasdaq Nordic and the New York Stock Exchange. She holds a PhD in industrial relations from the London School of Economics, where her research focused on organizational development and transformational change. She is an active voice in industry forums and events, and has authored articles and books about the topic, including: Service Leadership. The Quest for Competitive Advantage (Grönfeldt and Strother, 2006).

Svafa's work centers on understanding and improving the dynamic relationship between people and their environments. Drawing on her industry experience and lifelong passion for education and design, she continues to drive the development of new theories and practices in the emerging field of design innovation.

Sigurdur Thorsteinsson

Co-Founder Design Group Italia
Chief Design Officer Blue Lagoon

Sigurður Þorsteinsson is the Chief Brand, Design, and Development Officer at Blue Lagoon Iceland. Since 1997, he has overseen all strategic design initiatives ranging across scales, including products, experiences, and facilities. His leadership has been instrumental in shaping the Blue Lagoon transformation into a globally recognized well-being brand.

In parallel, Sigurður served as partner and chief design director at Design Group Italia (DGI), a Milan-based design and innovation firm, until 2023. During his three decades at DGI, he led complex, cross-sectoral projects ranging from health care and engineering to consumer goods, food systems, and destination experiences. His clients have included Samsung, 3M, ABB, Kone, Pepsico, Unilever, Barilla, Chicco, Flos, B&B Italia, and Moleskine, to name only a few.

Sigurður's practice moves beyond commercial design into visionary, purpose-driven collaborations, —including Björk's Náttúra initiative and Vatnavinir, a transdisciplinary think tank exploring water and wellbeing.

He has partnered with UNICEF to rethink fundraising strategies and with the United Nations Industrial Development Organization (UNIDO) on design-led approaches to economic development. As an educator, he has taught at Politecnico di Milano, Istituto Europeo di Design (IED) in Milan, and the Iceland University of the Arts, guiding emerging designers at the intersection of design theory, innovation, and societal transformation.

Sigurður has won numerous awards, such as the Compasso d'Oro, The Good Design, iF, Ahead, and multiple Red Dot Awards, including Red Dot Best-of-the-Best. His work is regularly featured in international publications such as *Wallpaper*, *Dezeen*, *Time*, and the *Financial Times*.

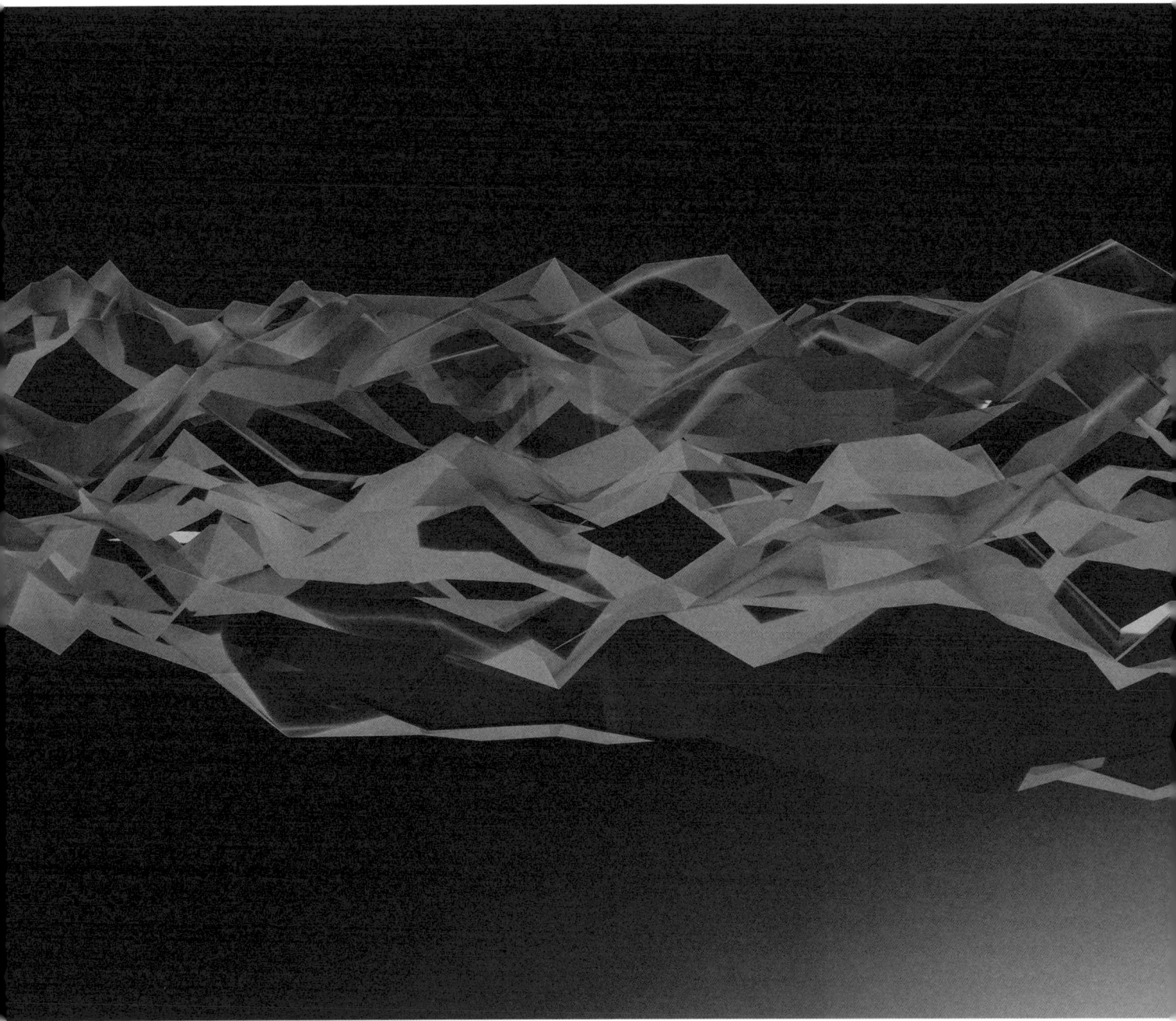

Appendix

Methods   175

Contributors   176

List of Interviewees   178

List of Images   187

References   191

Methods

This project set out to understand the power of design to affect the human experience and how design can be applied beyond its traditional boundaries. The book itself is a design experiment shaped through iterations, dialogue, and real-world application across industries and academia. The ideas and concepts presented are based on the authors' experience as well as our interpretation of data collected through interviews, review of evolving definitions of design, and explorations of moments in history where design has played a pivotal role in changing our view of the world and humans' place within it.

Semi-structured, in-depth interviews explored how individuals understand and apply design in diverse, real-world contexts. This interpretive approach allowed us to capture the complexity of lived experiences and uncover deeper insights that quantitative methods might overlook.

Following a theoretical sampling approach, 67 interviewees were selected for their potential to provide diverse perspectives on the application of design in complex environments, spanning multiple industries and geographies, including Africa, Asia, Europe, Latin America, and the United States. Participants represented a range of disciplines, including technology, health care, finance, public policy, education, energy, mobility, media, and social innovation. (See the Appendix for a full list of interviewees.)

Interview transcripts were analyzed using AI tools (atlasti. com and grain.com) to organize the data and generate codes based on contextual and semantic data patterns. These codes capture key concepts and reflect how each was ranked according to both frequency and contextual significance, factoring in not just how often a concept was mentioned, but also the depth of discussion and its relationship to other concepts. This approach surfaces concepts that are most meaningfully embedded in the data, offering a more nuanced view than frequency alone. Using an inductive approach, codes emerged without a predefined framework. These AI-generated codes were then reviewed, refined, and the authors' interpretation of the results is presented in the book.

Contributors

María Esteban Casañas
Architect and educator, A.I., innovation, architectural design, and computation
Project Lead, MITdesignX, Massachusetts Institute of Technology
Programa de Doctorado en Arquitectura y Urbanismo, Universidad Politécnica de Madrid

Matilda Bathurst
Development Editor

Gianandrea Giacoma
Psychologist, psychotherapist, and consultant

Francesco Zurlo
Strategic, systematic designer and educator
Dean of the School of Design, Politecnico di Milano
Professor of Industrial Design, Politecnico di Milano
Director of the Design + Strategies research group, Politecnico di Milano

Dale Dillavou
Psychologist, science advisor

Luca De Biase
Journalist and author
Founder and chief editor, Nòva

Research Assistants

Yasuyuki Hayama
Strategic designer and educator
Assistant Professor, Faculty of Design, Kyushu University

Gilad Rosenzweig
Architect and urban planner
Executive Director of DesignX, MIT SA+P

Carla Sedini
Researcher and educator, sociology and design
Associate Professor at IULM University, Faculty of Arts, Fashion and Tourism

Michael Stradley
Designer and educator
Ventulett NEXT Fellow at Georgia Tech School of Architecture

List of Interviewees

1   **'Kayode Agbalajobi**
Real Estate Asset Manager
Director, Asset Management & Sustainability at Carr Properties
Interview date: 11 August 2021

2   **Azra Aksamija**
Artist, architectural historian
Director of the Art, Culture and Technology Program
Director of the Future Heritage Lab
Interview date: 19 April 2025

3   **Ricardo Álvarez**
Technology Innovator, urban creative
Postdoctoral Fellow at the MIT Senseable City Lab
Interview date: 5 August 2021

4   **Masayo Ave**
Designer, sensory experience
Founder, MasayoAve creation
Interview date: 5 August 2021

5   **Tomoko Azumi**
Interior designer, products, furniture and space
Director, TNA Design Studio
Interview date: 19 August 2021

6   **Gloria Barcellini**
Architect
Metadesign, Alessi
Interview date: 10 May 2021

7   **Alberto Bassi**
Design historian and critic
Director of graduate program in design of the Università Iuav di Venezia
Interview date: 11 May 2021

8   **Kevin Bethune**
Author, design innovator
Founder & Chief Creative Officer of dreams • design + life
Interview date: 14 October 2022

9   **Alfredo Biffi**
Leadership, human resources and digital technologies
Associate Professor of Business Organization, Economics Department, Insubria University (Varese)
Professor of Business Organization, Università Cattolica del Sacro Cuore (Milan)
Affiliate Professor of Leadership, Human Resource and Digital Technologies SDA Bocconi School of Management (Bocconi University, Milan).
Interview date: 29 September 2022

10   **Sigrún Birgisdóttir**
Architect, educator
Former dean of Design and Architecture, Iceland Academy of the Arts
Interview date: 2 September 2021

11   **Luisa Bocchietto**
Architect and industrial designer
Former President of the Industrial Design Association, Italy
Senator, World Design Organization
Interview date: 14 May 2021

12   **Michele Boldrin**
Complexity and human behavior
Joseph Gibson Hoyt Distinguished Professor in Arts & Sciences, Washington University in Saint Louis
Co-Director of the Center for Dynamic Economics, Washington University in Saint Louis
Interview date: 26 October 2022

13   **Sofía Bosch Gómez**
Designer and researcher, social change
Assistant Professor, Northeastern's College of Arts, Media, and Design.
Interview date: 15 June 2021

14   **Marco Carvalho**
Computer Science, machine learning and data mining
Professor in Computer Sciences, Florida Institute of Technology
Executive Director, L3Harris Institute for Assured Information
Interview date: 13 December 2022

15   **Jeffrey Cassis**
Executive leader, electronic manufacturing, software and AI
Former CEO, Philips Color Kinetics
Interview date: 21 October 2022

16   Andrea Chegut
Innovator, research scientist, and financial econometrician
Founder and Director of the MIT Real Estate Innovation Lab
Head of Research and co-founder of MITdesignX
Interview date: 8 June 2021

17   Elizabeth Christoforetti
Systems designer and technologist
Assistant Professor in Practice of Architecture, Harvard GSD
Founding Principal, Supernormal
Interview date: 27 May 2021

18   Matthew Claudel
Urban designer
Founder, Field States
Interview date: 12 May 2021

19   Marcelo Coelho
Computation artist and designer
Director, MIT Design Intelligence Lab
Interview date: 21 June 2021

20   Peter Coughlan
Strategic designer and systems innovator
Co-founder of IDEO's Transformation by Design practice
Interview date: 13 April 2021

21   Tim Cousin
Architect
Co-founder, Roofscapes
Interview date: 30 June 2022

22   Alessandro Cravera
Management innovation
Senior Partner, Newton S.p.A.
Faculty Member ALTIS Università Cattolica
Coordinator and lecturer, Leadership & Management Development Laboratory, Executive MBA, 24 Ore Business School
Faculty Member CASD (Centro Alti Studi della Difesa)
Interview date: 29 September 2022

23   Luisa Damiano
Epistemology of complex systems; Philosophy of emerging sciences and technologies
Professor of Logic and Philosophy of science, IULM University, Milan
Interview date: October 31, 2022

24   Nicholas de Monchaux
Designer, author, educator
Professor and Head of Architecture at MIT
Interview date: 15 September 2021

25   **Olivier Faber**
Architect
Co-founder, Roofscapes
Interview date: 30 June 2022

26   **Rubén Fernández-Costa O'Dogherty**
Journalist and writer, information sciences
Journalist, El Español, Forbes, Expansión, Vogue, Harper's Bazaar
Professor and researcher, Universidad de Nebrija, Universidad Complutense
Interview date: 16 September 2021

27   **Daniel Fink**
Computational design, property development, and geospatial science
Partner, Black Oak Group
Interview date: 24 May 2021

28   **Luca Foresti**
Financial mathematics, physics, and complexity management
CEO of Centro Medico Santagostino, Società e Salute SPA
CEO and Founder, First Principles SRL
Chairman of the Board and Co-Founder, Run2AI
Interview date: 18 October 2022

29   **Dai Fujiwara**
Designer, creative director
Founder, DAIFUJIWARA AND COMPANY, DAI&Co.
Interview date: 26 Aug 2021

30   **Luciano Galimberti**
Designer
President, ADI Associazione per il Disegno Industriale
Co-Founder and Design Manager, BGPIU' Studio
Interview date: 14 June 2021

31   **Tony Hu**
Electrical engineer and product designer
Program Director and Senior Lecturer of the Riccio Graduate Engineering Leadership Program
Interview date: 16 September 2021

32   **Caroline Jones**
Art historian, curator, author, critic
Rudge (1948) and Nancy Allen Professor in the History, Theory, and Criticism, Department of Architecture,
Associate dean, MIT SA+P
Interview date: 1 June 2021

33   **Kawasaki, Kazuo**
Industrial designer
Professor at Osaka University
Interview date: 24 August 2021

34   **Jeff Klug**
Architect and educator
Former Director, Design Discovery, Harvard GSD
Interview date: 6 August 2021

35   **Tamara Knox**
Real estate developer, city design and development
CEO and Co-Founder, Frolic Community
Interview date: 1 September 2022

36   **Kengo Kuma**
Architect
Kengo Kuma and Associates
Architect, University Professor and Professor Emeritus at the University of Tokyo
Interview date: 18 August 2021

37   **Eytan Levi**
Architect
Co-founder, Roofscapes
Interview date: 30 June 2022

38   **Halla Hrund Logadóttir**
Scientist and politician
Co-founder and director, Arctic Initiative at Harvard Kennedy School
Founder, Arctic Innovation Lab
Director-General, Iceland's National Energy
Interview date: 23 August 2022

39   **Antonio Macchi Cassia**
Designer
Founder, Macchi Cassia Studio
Interview date: 11 May 2021

40   **Ezio Manzini**
Design academic and author, social innovation and sustainability
Honorary Professor, Politecnico di Milano
Chair Professor at University of the Arts London
Distinguished Professor on Design for Social Innovation at ELISAVA
Interview date: 01 June 2021

41   **Melissa Marsh**
Workplace strategist, change management
Founder & CEO of PLASTARC
Interview date: 16 June 2021

42   **Stefano Mirti**
Architect, educator, strategic design
Founder and head of strategy, IdLab
Interview date: 13 July 2021

43 Chiara Montanari
Engineer, explorer, Antarctic Mindset Trainer
Founder, Antarctic Mindset
Former Head of Expedition to Antarctica (First Zero Emission Research Station)
Interview date: 19 September 2022

44 Joshua Morrison
Urban Designer, city design and development
Co-Founder and Chief Operating Officer, Frolic Community
Interview date: 1 September 2022

45 Takehiko Nagakura
Architect and educator
Founder, Architecture, Representation and Computation group (ARC)
Professor, School of Architecture and Planning, Massachusetts Institute of Technology
Interview date: 16 September 2021

46 Dava Newman
Aerospace biomedical engineer
Director of the MIT Media Lab
Apollo Professor of Astronautics
Former deputy administrator of NASA
Interview date: 26 August 2021

47 Don Norman
Researcher, professor, and author
Director of The Design Lab, University of California, San Diego
Co-founder, Nielsen Norman Group
Professor Emeritus of Cognitive Science, University of California, San Diego
Distinguished Visiting Professor, Korea Advanced Institute of Science and Technology
Interview date: 2 June 2021

48 John Ochsendorf
Engineer, designer, educator
Professor, dual appointment in the Department of Architecture and the Department of Civil and Environmental Engineering at MIT
Founding Director MIT Morningside Academy for Design (MAD)
Interview date: 10 August 2022

49 Francesco Pagnini
Researcher, mind-body connection, clinical psychology, health psychology, mindfulness
Professor of Clinical Psychology, Department of Psychology, Università Cattolica del Sacro Cuore
Interview date: 3 April 2023

50 Anty Pansera
Art and Design historian
Founder, Studio Pansera
Interview date: 29 June 2021

51 Bill Patrowicz
Strategic advisor and innovator
Principal, Kaiser Research
Executive Counselor, Arctic Center of Excellence, U.S. Department of Homeland Security
Interview date: 6 December 2022

52 Jim Peraino
Software engineer and architect, public health, evaluation, and behavioral economics
Technical Director, product engineering, Norm Ai
Interview date: 12 May 2021

53 Paul Pettigrew
Architect, relationships between cities, buildings and functional objects
Founder, Paul Pettigrew Architect & Products in Space
Interview date: 12 May 2021

54 Ambrogio Rossari
Industrial designer
Founder, Rossari & Associati studio
Interview date: 29 June 2021

55 Taku Satoh
Graphic designer
Founder, TSDO
Interview date: 17 August 2021

56 Lisbeth Shepherd
Social innovator
Lecturer, MIT School of Architecture
Entrepreneur in Residence, MITdesignX
Interview date: 7 September 2021

57 Molly Wright Steenson
Historian of Artificial Intelligence
President and CEO of the American Swedish Institute
Interview date: 24 May 2021

58 Skylar Tibbits
Designer, computer scientist
Associate Professor of Design Research in the Department of Architecture
Founder, MIT's Self-Assembly Lab
MAD's assistant director for education
Interview date: 13 September 2024

59 Pietro Trabucchi
Psychologist, resilience, motivation and stress management
Lecturer, University of Verona, Department of Neurological, Neuropsychological, Morphological and Movement Sciences
Research Lead on the psychology of activities in extreme environments at CeRiSM, the Verona-Trento Interuniversity Research Center on human performance
Interview date: 28 October 2022

60 **Laura Traldi**
Design journalist
Journalist and editor, Interni Magazine
Strategic communications consultant
Interview date: 6 July 2021

61 **Clino Trini Castelli**
Industrial designer and artist
Founder, Castelli Design
Founder, Domus Academy
Interview date: 18 May 2021

62 **Nathalie van Bockstaele**
Researcher and choreographer of speech and action
Independent researcher in human interactions and non-verbal thinking
Interview date: 27 August 2021

63 **Roberto Verganti**
Scholar and author
Professor, The Josefsson Family Chair in Art and Innovation,
Director, Center for Art and Innovation, Stockholm School of Economics
Visiting Lecturer, Harvard Business School
Interview date: 8 September 2021

64 **Maria Paulina Villa Posada**
Architect and urban planner, knowledge economy
Portfolio Manager, Corporación Ruta N
Interview date: 10 August 2021

65 **Rachelle Villalon**
Designer, computational design
Founder, Executive Chairman & Chief Science Officer, Hosta AI
Interview date: 17 August 2022

66 **Shunji Yamanaka**
Design engineer, mechanical and biofunctional systems
Professor at the University of Tokyo
Interview date: 30 July 2021

67 **Maria Yang**
Mechanical engineer and product designer
Deputy Dean, MIT Engineering
Interview date: 16 August 2021

List of Images

Page	Image	Source
Endpapers	Spinning fractal burst on black background	Adobe Stock, Primada
16	Aerial view of highway junctions shape letter x cross at night, Kuala Lumpur.	Adobe Stock, tampatra
19	White clouds as an x shape	Adobe Stock, Harry Wedzinga
20	A volcanic eruption by the Blue Lagoon	Courtesy of Blue Lagoon Iceland
23	Barrett Lyon "The Opte Project"	Courtesy Barret Lyon
28	People inside the [Guggenheim] Museum	Pexels, Malcolm Hill
30	Architectural photography of glass building	Pexels, Soloman Soh
37	White concrete stairway	Pexels, Jan van der Wolf
39	Reflection in mirror of people walking stairs	Pexels, Soner Arkan
41	A rendering of a housing solution	Courtesy of Frolic Community
42	Rosetta Stone	Adobe Stock, Jens Teichman
45	Black and white photography of a blurred reflection of a person	Pexels, Beyza Kaplan
46	Blombos Cave, drawing on tool	Stephen Alvarez, Alvarez Photography
46	Aristotle	Public domain. Wikimedia Commons
46	Piazza del Popolo, Rome	Luigi Zomparelli
47	Vaucanson's Automatic Duck	Public domain. Wikimedia Commons
47	Ten year old spinner in a North Carolina cotton mill, captured by Lewis W. Hine (1900 - 1937)	Photography Collection, New York Public Library
47	Marcel Breuer's Wassily Chair	Marcel Breuer
47	AI-generated content. Neural network synapse activity in human brain illustration	Pixtastock. Anton Gvozdikov
48	Blombos Cave, drawing on tool	Stephen Alvarez, Alvarez Photography
48	Homo naledi symbols in Rising Star Cave, South Africa	Lee Berger, Cave of Bones
49	Aristotle	Public domain. Wikimedia Commons
50	Medieval depiction of the Last Supper.	Public domain. Wikimedia Commons
50	The Last Supper by Leonardo DaVinci	Public domain. Wikimedia Commons
51	Piazza del Popolo, Rome	Luigi Zomparelli
51	Diagram demonstrating Filippo Brunelleschi's perspective technique	Scala / Art resource NY

Page	Image	Source
52	Vaucanson's Automatic Duck	Public domain. Wikimedia Commons.
54	Early utopian Lowell, Massachusetts view, 1838.	Britannica ImageQuest, Encyclopædia Britannica
54	The Lowell Offering Masthead, Mill Girls in 19th century print	The Lowell Offering
55	Boott Cotton Mills workers in mill yard, c. 1884	Public domain. Lowell Historical Society
55	Ten year old spinner in a North Carolina cotton mill, captured by Lewis W. Hine (1900 - 1937)	Photography Collection, New York Public Library
56	Daniel Hudson Burnham, View of the proposed Civic Center Plaza and Buildings, The 1909 plan of Chicago, 1908	Public domain. The Art Institute of Chicago
57	The Bauhaus in Dessau	Andreas Meichsner for New York Times
57	Marcel Breuer's Wassily Chair	Marcel Breuer
58	Spectators marvel at the Futurama model at the 1939 New York's World's Fair.	General Motors
58	Futurama, New York World's Fair 1939, designed by Norman Bel Geddes. Photograph by Richard Garrison	Richard Garrison
59	AI-generated content. Neural network synapse activity in human brain illustration	Pixtastock. Anton Gvozdikov
61	Scientists inspect a pothole	Courtesy of Biobot
62	Architect working on blueprint	Adobe Stock, Meow Creations
65	"A Walk with the Child", Paul Klee	University of Michigan Museum of Art
69	Construction of the Central Artery, Boston, 1950s	City, Territory and Architecture Journal
70	Figure-ground diagram: a vase or two faces?	Public domain. Wikimedia Commons
71	Boulevard Upper Broadway ca. 1880	Bridgeman Images
71	Upper Broadway today, 2025	Google Earth
72	Mathematical proportions of the human head. Leonardo da Vinci, 1490	Public domain. Wikimedia Commons
74	A digital circuit board pattern, generated with AI	Adobe Stock, KEOCHAN
76	Modern architectural composition with shadows	Pexels, Tiến Nguyễn
78	ISS over the planet Earth. Elements of this image furnished by NASA	Adobe Stock, Artisom P
80	Macro leaves background texture	Adobe Stock, Valentina R

Page	Image	Source
83	Tiktaalik, extinct walking fish	Adobe Stock, dottedyeti
85	Close up of ice	Pexels, Batın Özen
87	A woman with deep fried pastries	Courtesy of Quipu
88	Abstract Icelandic glacier rivers melting pattern in summer	Adobe Stock, Mumemories
91	A swirling tornado of Barracuda	Adobe Stock, whitecomberd
98	Pattern against dark background	Pexels, Ramon Karolan
107	Film negatives arranged in overlapping layers	Adobe Stock, Jumpingsack
109	People on stairs	Pexels, Sasha P
110	Blurred woman standing by table	Pexels, cottonbro studio
113	High lighthouse stairs, Vierge Island, Brittany, France	Adobe Stock, WIlly Mobilo
117	A rendering of rooftop gardens in Paris	Courtesy of Roofscapes
118	Lava flowing along the defense wall surrounding the Blue Lagoon, Iceland	Blue Lagoon. Cindy Rún Li
119	Blue Lagoon ecocylce	Courtesy of Blue Lagoon. Igor Micevic
120	Aerial view of the Blue Lagoon facilities in winter	Blue Lagoon. Cindy Rún Li
121	The Svartsengi power plant as seen from The Blue Lagoon ca 2013	Courtesy of Blue Lagoon
121	Blue Lagoon patrons do exercises in the lagoon	Courtesy of Blue Lagoon
122	Exterior view of the Retreat Spa Restaurant	Blue Lagoon. Ari Magg
123	Blurring the lines between inside and outside. Lava Restaurant at the Blue Lagoon	Blue Lagoon. Cindy Rún Li
123	Glass vases and bowl sit on top of a table made from lava from the premises	Blue Lagoon. Giorgio Possenti
123	A view from a Retreat suite overlooking the moss-covered lava	Blue Lagoon. Giorgio Possenti
124	Early Blue Lagoon bathing facility, 1992	Courtesy of Blue Lagoon
124	Guests enjoy the lagoon, with the Svartsengi power plant in the background	Courtesy of Blue Lagoon
124	R&D lab scientists researches the unique blue-green algae found in the lagoon water	Blue Lagoon. Cindy Rún Li
125	Blue Lagoon skincare products	Blue Lagoon. Cindy Rún Li
125	Blue Lagoon guests with silica mud face mask	Blue Lagoon. Cindy Rún Li
125	An exterior view of the Retreat hotel	Blue Lagoon. Giorgio Possenti

Page	Image	Source
126	Simone Young conducts Richard Strauss	Craig Abercrombie, The Sydney Symphony Orchestra
129	View from below of a silhouette of a person swimming in a pool with glass floor	Adobe Stock, victor
134	Young girl dancing. Long exposure. Contemporary hip hop dance	Adobe Stock, Georgii
137	Low angle photo grayscale of person tightrope walking	Pexels, Marcelo Moreira
138	Aerial view of two people walking on the field	Pexels, Valter Zhara
140	Blurred reflection in antique mirror display	Pexels, Juan manuel Perez
145	Learning Beautiful education set	Courtesy of Learning Beautiful
146	Modern bridge surrounded by fog	Adobe Stock, Heath Canjandig/Wirestock
149	Black and white photo of the sun shining through the railing	Pexels, Juan Fam
150	Aerial view of person in yellow raincoat walking across rice field	Pexels, Esra Bürçün
152	Arcaden [Breakwater columns in Tazacorte, Canary Islands]	Adobe Stock, Denis
154	White paper boat floating on water	Pexels, Pixabay
156	Bird's eye view forest and city border Vila Caiçara, Praia Grande, Sao Paulo, Brazil	Adobe Stock, gokturk_06
158	Floor tiles	Pexels, sum+it
162	Metal printing press letters	Adobe Stock, zlikovec
164	Scenography of dancers dancing	Adobe Stock, Giovanni Nitti
165	Scenography of dancers dancing	Adobe Stock, Giovanni Nitti
167	Indian women interacting with Bandhu app on mobile devices	Courtesy of Bandhu
168	Ammonite, prehistoric fossilized mollusk, an extinct marine animal	Pexels, adrian vieriu
170	The authors from left to right: S. Thorsteinsson. S. Grönfeldt, D. Frenchman	María Esteban Casañas
172	Gray texture structure	Pexels, Steve Johnson
174	Brown textile [External façade of Mo POP Museum in Seattle]	Pexels, Scott Webb
176	Large crowd of people walking in city street during daytime, view from above.	Adobe Stock, eartist85
190	Water dripping from tile surface	Pexels, Rachel Claire
Endpapers	Spinning fractal burst on black background	Adobe Stock, Primada

References

AARP. (2021). *Where we live, where we age: Trends in home and community preferences.* https://datastories.aarp.org/2021/home-and-community-preferences/

African Development Bank Group. (2023). *African economic outlook 2023.* AfDB Group Publications.

Al-Khalili, J. (2011). *The house of wisdom: How Arabic science saved ancient knowledge and gave us the Renaissance.* Penguin Press.

Albert, D. Z. (1992). *Quantum mechanics and experience.* Harvard University Press.

Alberti, L. B. (1991). *On painting ("De Pictura"*, C. Grayson, Trans.). Penguin Classics. (Original work published 1435.)

Alighieri, D. (2002). *The divine comedy* (J. Ciardi, Trans.). New American Library. (Original work published ca. 1320.)

Baclawski, K. (2018). *The observer effect.* Retrieved from https://www.researchgate.net/publication/326795653_The_Observer_Effect

Barnett, R. (2000). *Realizing the university in an age of supercomplexity.* Open University Press.

Bel Geddes. N. (1939). "Highways and Horizons" (Voice-over exhibit narration). General Motors Futurama. (1939-40). To new horizons [Film]. Jam Handy Organization. https://www.youtube.com/watch?v=sClZqfnWqmc

Bellamy, E. (2009). *Looking backward: 2000–1887.* Oxford University Press. (Original work published 1888.)

Bennett, N., & Lemoine, G. J. (2014). What VUCA really means for you. *Harvard Business Review*, 92(1/2), 27.

Berger, L. R., Hawks, J., Fuentes, A., van Rooyen, D., Tsikoane, M., Ramalepa, M., Nkwe, S., & Molopyane, K. (2023, June 5). 241,000 to 335,000 years old rock engravings made by Homo naledi in the Rising Star Cave Systems, South Africa. *BioRxiv.* https://doi.org/10.1101/2023.06.01.543133

Bethune, K. G. (2024). *Reimagining design: unlocking strategic innovation.* MIT Press.

Born, M. (1971). *The Born-Einstein Letters: Correspondence between Albert Einstein and Max and Hedwig Born from 1916 to 1955* (I. Born, Trans.). Walker and Company.

Brookfield, S. D. (2009). The concept of critically reflective practice. In N. Lyons (Ed.), Handbook of reflection and reflective inquiry: Mapping a way of knowing for professional reflective inquiry (pp. 127–148). Springer.

Brown, B. (2018). *Dare to lead: Brave work. Tough conversations. Whole hearts.* Random House.

Caporale, L. H. (2003). Natural selection and the emergence of a mutation phenotype: an update of the evolutionary synthesis considering mechanisms that affect genome variation. *Annual Review of Microbiology, 57,* 467–485. https://doi.org/10.1146/annurev.micro.57.030502.090855

Chomsky, N., & Moro, A. (2022). *The secrets of words.* MIT Press.

Clark, A. (2024). *The experience machine: How our minds predict and shape reality.* Random House.

Code.org, CSTA, & ECEP Alliance (2022). 2022 State of Computer Science Education: *Understanding Our National Imperative.* https://advocacy.code.org/stateofcs

Cramer, J. G. (2019). Quantum entanglement across time. *Analog: Science Fiction & Fact Magazine.* https://www.npl.washington.edu/av/altvw203.html.

Crawford, M. (1995). *Building the workingman's paradise: The design of American company towns.* Verso.

Csikszentmihalyi, M. (1990). *Flow: The psychology of optimal experience.* Harper & Row.

Darwin, C. (1869). *On the origin of species by means of natural selection, or the preservation of favoured races in the struggle for life* (5th ed.). John Murray.

Deacon, T. W. (2012). *Incomplete nature: How the mind emerged from matter.* W. W. Norton Company.

Deming, D. J., & Noray, K. (2020). Earnings dynamics, changing job skills, and STEM careers. *The Quarterly Journal of Economics, 135*(4), 1965–2005. https://doi.org/10.1093/qje/qjaa021

Descartes, R. (1995). *Discourse on the method of rightly conducting the reason, and seeking truth in the sciences.* (J. Veitch, transl.) Project Gutenberg ebook #57. (Original work published 1637.) https://www.gutenberg.org/files/59/59-h/59-h.htm

Doudna, J. A., & Charpentier, E. (2014). The new frontier of genome engineering with CRISPR-Cas9. *Science, 346*(6213), 1258096. https://doi.org/10.1126/science.1258096

Eberly, R. A. (2022). Flesh became words. *MIT Technology Review.* https://www.technologyreview.com/2022/04/27/1048493/flesh-became-words/

Edgerton, S. Y. (2006). Brunelleschi's mirror, Alberti's window, and Galileo's 'perspective tube'. *História Ciências Saúde-Manguinhos, 13*(suppl), 151–179. https://www.scielo.br/j/hcsm/a/ZKBpG6VdWvNmKPpnXsP8nBq/?lang=en

Edgerton, S. Y. (2010). Space, vision, and faith: Linear perspective in Renaissance art and science. *Nexus Network Journal, 12*(1), 13–30. https://doi.org/10.1007/s00004-010-0056-7

Evans, A. B. (2013). Jules Verne's vision of the future city. Paris in the twentieth century. *Science Fiction Studies, 40*(2), 240–255.

Fournier, G. P., Moore, K. R., Rangel, L. T., Payette, J. G., Momper, L., & Bosak, T. (2021). The Archean origin of oxygenic photosynthesis and extant cyanobacterial lineages. *Proceedings of the Royal Society B: Biological Sciences, 288*(1959), 20210675. https://doi.org/10.1098/rspb.2021.0675

Freeman W. J. (2009). Vortices in brain activity: their mechanism and significance for perception. *Neural Networks, 22*(5-6), 491–501. https://doi.org/10.1016/j.neunet.2009.06.050

Frenchman, D. (1990). Narrative places and the new practice of urban design. In L. Vale & M. Warner (Eds.), *Imaging the city: Continuing struggles and new directions* (pp. 257–282). Routledge.

Frenchman, D. (1995). Cities of Tomorrow [Course Lecture 11.335j/4.748j]. Massachusetts Institute of Technology.

Gabor, D. (1963). *Inventing the future.* Penguin/Secker & Warburg. http://ci.nii.ac.jp/ncid/BA19121296

Gale, J., Alemdar, M., Lingle, J., & Newton, S. (2020). Exploring critical components of an integrated STEM curriculum: An application of the innovation implementation framework. *International Journal of STEM Education, 7,* 5.

Gasparrini, A., Guo, Y., Sera, F., Vicedo-Cabrera, A. M., Huber, V., Tong, S., de Sousa Zanotti Stagliorio Coelho, M., Nascimento Saldiva, P. H., Lavigne, E., Matus Correa, P., Valdes Ortega, N., Kan, H., Osorio, S., Kyselý, J., Urban, A., Jaakkola, J. J. K., Ryti, N. R. I., Pascal, M., Goodman, P. G., Zeka, A., … Armstrong, B. (2017). Projections of temperature-related excess mortality under climate change scenarios. *The Lancet Planetary Health, 1*(9), e360–e367. https://doi.org/10.1016/S2542-5196(17)30156-0

Gattupalli, A. (2024, September 16). How Roofscapes' Paris pilot project is pioneering climate-resilient architecture in Europe. *Arch Daily.* https://www.archdaily.com/1021038/how-roofscapes-paris-pilot-project-is-pioneering-climate-resilient-architecture-in-europe

Gobbo, F. (2011). Using ethnography to explore culture in education: An example from intercultural research. In S. Delamont (Ed.), *Handbook of qualitative research in education* (pp. 232–243). Edward Elgar Publishing.

Griffin, A., & Williams, M. (2024). Screen time guidelines for kids, at every age: CHLA experts weigh in. Children's Hospital Los Angeles. https://www.chla.org/blog/advice-experts/screen-time-guidelines-kids-every-age-chla-experts-weigh

Grönfeldt, S. (2003). *Customer oriented behavior, nature, impact and development.* Reykjavik University Press.

Grönfeldt, S., & Strother, J. (2006). *Service leadership: The quest for competitive advantage.* Sage.

Hallström, J., & Ankiewicz, P. (2023). Design as the basis for integrated STEM education: A philosophical framework. *Frontiers in Education, 8,* 1–8.

Harrison, E. (1986). Newton and the infinite universe. *Physics Today, 39*(2), 24–32. https://doi.org/10.1063/1.881049

Hassell, J. M., Begon, M., Ward, M. J., & Fèvre, E. M. (2017). Urbanization and disease emergence: Dynamics at the wildlife-livestock-human interface. *Trends in Ecology & Evolution, 32*(1), 55–67. https://doi.org/10.1016/j.tree.2016.09.012

International Labour Organization (ILO). (2023a). Charting progress on the global goals and decent work. ILOSTAT. https://ilostat.ilo.org/blog/charting-progress-on-the-global-goals-and-decent-work

International Labour Organization. (2023b). Statistics on the informal economy. ILOSTAT. https://ilostat.ilo.org/topics/informality/

Johnson, P., & Hitchcock, H. R. (1932). *The international style*. Museum of Modern Art.

Kaushik, K., & Campbell, J. (2023). India's migrant millions: Caught between jobless villages and city hazards. Reuters. https://www.reuters.com/world/india/indias-migrant-millions-caught-between-jobless-villages-city-hazards-2023-04-18/

Iskander, N. (2018). Design thinking is fundamentally conservative and preserves the status quo. *Harvard Business Review*. https://hbr.org/2018/09/design-thinking-is-fundamentally-conservative-and-preserves-the-status-quo

Kárpáti, Z. (2019). Michelangelo, the David, and Donatello. In Z. Kárpáti, E. Nagy, & P. Ujvári (Eds.). *Triumph of the body: Michelangelo and sixteenth-century Italian draughtsmanship*. Museum of Fine Arts, Budapest. https://www.academia.edu/38784064/Triumph_of_the_Body_Michelangelo_and_Sixteenth_Century_Italian_Draughtsmanship

Keil, G., & Kreft, N. (Eds.). (2019). Human beings as rational animals. *In Aristotle's Anthropology* (pp. 23–96). Cambridge University Press.

Kemp, M. (1990). *The science of art: Optical themes in western art from Brunelleschi to Seurat*. Yale University Press.

Kemsley, J. (2019, July 16). Behind the scenes of the STEM-humanities culture war. *Chemical & Engineering News, 97*(29). https://cen.acs.org/education/undergraduate-education/Behind-the-scenes-STEM-humanities-culture-war/97/i29.

Klee, P. (1953). *Pedagogical sketchbook*. [Introduction and translation by S. Moholy-Nagy]. F. A. Praeger.

Kostof, S. (1991). *The city shaped: Urban patterns and meanings through history*. Thames & Hudson.

Le Corbusier. (1967). *The radiant city (La Ville Radieuse): Elements of a doctrine of urbanism to be used as the basis of our machine-age civilization* (P. Knight, Trans.). Orion Press. (Original work published 1933.)

Lesso, R. (2022, October 11). Who discovered linear perspective? https://www.thecollector.com/who-discovered-linear-perspective/

Lindow, J. (2002). *Norse mythology: A guide to gods, heroes, rituals, and beliefs*. Oxford University Press.

Mandt, R., Seetharam, K., & Chang, C. H. M. (2020). Federal R&D funding: The bedrock of national innovation. *Science Policy Review, 1*, 44–54. https://doi.org/10.38105/spr.n463z4t1u8

May, J. (2021). *Bias in science: Natural and social.* Synthese, 199, 3345–3366. https://doi.org/10.1007/s11229-020-02722-6

Masselot, P., Mistry, M., Vanoli, J., Schneider, R., Iungman, T., Garcia-Leon, D., Ciscar, J. C., Feyen, L., Orru, H., Urban, A., Breitner, S., Huber, V., Schneider, A., Samoli, E., Stafoggia, M., de'Donato, F., Rao, S., Armstrong, B., Nieuwenhuijsen, M., Vicedo-Cabrera, A. M., … EXHAUSTION project (2023). Excess mortality attributed to heat and cold: a health impact assessment study in 854 cities in Europe. *The Lancet: Planetary Health, 7*(4), e271–e281. https://doi.org/10.1016/S2542-5196(23)00023-2

McGilchrist, I. (2019). *The master and his emissary: The divided brain and the making of the western world* (2nd ed.). Yale University Press.

McGilchrist, I. (2021). *The matter with things: Our brains, our delusions and the unmaking of the world* (vol. I and II). Perspectiva Press.

McAuliffe, M., & Oucho, L. A. (Eds.). (2024). *World migration report 2024.* International Organization for Migration (IOM). https://publications.iom.int/books/world-migration-report-2024

McKeown, G. (2014). *Essentialism: The disciplined pursuit of less.* Crown Business.

Meadows, D. H. (2012). *Leverage points: Places to intervene in a system.* https://donellameadows.org/archives/leverage-points-places-to-intervene-in-a-system/

Miller, G. A. (1956). The magical number seven, plus or minus two: Some limits on our capacity for processing information. *Psychological Review, 63*(2), 81–97. https://doi.org/10.1037/h0043158

Moddel, P. (2023). *The unified principle of colour.* Ibex Press.

Morowitz, H. J. (2002). *The emergence of everything: How the world became complex.* Oxford University Press.

Mullen, R. (2019, July 16). In a recent report reviewing evidence that "education programs integrating the arts, humanities, and sciences lead to improved educational and career outcomes." *Chemical & Engineering News, 97*(29). https://cen.acs.org/education/undergraduate-education/Behind-the-scenes-STEM-humanities-culture-war/97/i29

National Academy of Sciences. (n.d.). *Definitions of evolutionary terms.* Retrieved from https://www.nationalacademies.org/evolution/definitions on March 24, 2025.

Needham, J. (1995). *Science and civilisation in China.* Vol. 7, Part 2: General conclusions and reflections. Cambridge University Press.

Nielsen, J. A., Zielinski, B. A., Ferguson, M. A., Lainhart, J. E., & Anderson, J. S. (2013). An evaluation of the left-brain vs. right-brain hypothesis with resting state functional connectivity magnetic resonance imaging. *PLOS ONE, 8*(8), e71275. https://doi.org/10.1371/journal.pone.0071275

Norman, D. (2024). *Design for a better world*. MIT Press.

Norman, D. A. (2011, October 4). Design education: Brilliance without substance. *Core77*. https://jnd.org/design-education-brilliance-without-substance/

O'Connor, J. J., & Robertson, E. F. (1996, May). *A history of quantum mechanics*. University of St. Andrews. https://mathshistory.st-andrews.ac.uk/HistTopics/The_Quantum_age_begins/

Parry, R. (2024). Episteme and techne. In E. N. Zalta & U. Nodelman (Eds.), *The Stanford encyclopedia of philosophy* (Winter 2024 edition). https://plato.stanford.edu/archives/win2024/entries/episteme-techne

Pendleton-Julian, A. M., & Brown, J. S. (2018). Design unbound: *Designing for emergence in a white water world*. MIT Press.

Pisano, G. P. (2019). The hard truth about innovative cultures. *Harvard Business Review, 97*(1), 62–71.

Reece, B. C. (2018). Aristotle's four causes of action. *Australasian Journal of Philosophy, 96*(4), 794–809. https://doi.org/10.1080/00048402.2018.1482932

Resnick, B. (2020, October 26). 'Reality' is constructed by your brain. Here's what that means, and why it matters. *Vox*. https://www.vox.com/science-and-health/20978285/optical-illusion-science-humility-reality-polarization

Robinson, L., & Dewan, A. (2024, August 12). These cities will be too hot for the Olympics by 2050. CNN Climate. https://www.cnn.com/2024/08/11/climate/olympics-cities-extreme-heat/index.html

Schilling, D. R. (2013, April 19). *Knowledge doubling every 12 months, soon to be every 12 hours*. http://www.industrytap.com/knowledge-doubling-every-12-months-soon-to-beevery-12- hours/3950

Schön, D. (1984). *The reflective practitioner: How professionals think in action*. Basic Books.

Shapin, S. (2018). *The scientific revolution* (2nd ed.). University of Chicago Press.

Snow, J. (1855). *On the mode of communication of cholera* (2nd ed.). John Churchill.

Stang, N. F. (2024). Kant's transcendental idealism. In E. N. Zalta & U. Nodelman (Eds.), *The Stanford Encyclopedia of Philosophy* (Spring 2024 ed.). https://plato.stanford.edu/archives/spr2024/entries/kant-transcendental-idealism/

Statista. (2025, March 3). *Penetration rate of smartphones in Colombia from 2014 to 2029*. https://www.statista.com/forecasts/1143883/smartphone-penetration-forecast-in-colombia

The Bhagavad Gita (E. Easwaran, Trans., 2nd ed.). (2007). Nilgiri Press. (Original work published ca. 200 BCE.)

The epic of Gilgamesh: The Babylonian epic poem and other texts in Akkadian and Sumerian. (A. R. George, A. R. (Trans.). (2003). Penguin Classics. (Original work published ca. 2750–25100 BCE.)

United Nations Statistics Division (UNStats). (2023). *Sustainable Development Goals Report 2023: Goal 11 — Sustainable cities and communities*. https://ustats.un.org/sdgs/report/2023/goal-11

Verne, J. (1996). *Paris in the twentieth century (Paris au XXe siecle*; R. Howard, Trans.). Random House. (Original work created 1863; published in French, 1994).

Voiland, A. (2024, August 10). Running through Paris heat. NASA Earth Observatory. Retrieved March 16, 2025, from https://earthobservatory.nasa.gov

von Neumann, J. (1932). *The mathematical foundations of quantum mechanics*. Springer.

Wahba, S., & Ranarifidy, D. (2018, September 20). Re-awakening Kinshasa's splendor through targeted urban interventions. *The World Bank Blogs.* https://blogs.worldbank.org/africacan/re-awakening-kinshasas-splendor-through-targeted-urban-interventions

Wang, J. (2013). The importance of Aristotle to design thinking. *Design Issues, 29*(2), 4–15. http://www.jstor.org/stable/2426699

World Wildlife Fund. (2024). Living planet report: *A system in peril.* https://livingplanet.panda.org/en-GB/

For Further Reading

Aristotle. (2008). *Physics* (R. Waterfield, Trans.; D. Bostock, Ed.). Oxford University Press. (Original work published ca. 350 BCE.)

Bethune, K. G. (2025). *Nonlinear: Navigating design with curiosity and conviction.* The MIT Press.

Forrester, J. W. (1969). *Urban dynamics.* The MIT Press.

Graeber, D., & Wengrow, D. (2021). *The dawn of everything: A new history of humanity.* Farrar, Straus and Giroux.

Kaiser, D. (2020). *Quantum legacies: Dispatches from an uncertain world.* University of Chicago Press.

Lanza, R., & Pavšič, M., with Berman, B. (2020). *The grand biocentric design: How life creates reality.* BenBella Books

Noë, A. (2023). *The entanglement: How art and philosophy make us what we are.* Princeton University Press.

Ogas, O., & Gaddam, S. (2022). *Journey of the mind: How thinking emerged from chaos.* W. W. Norton & Company.

Panek, R. (2019). *The trouble with gravity: Solving the mystery beneath our feet.* Houghton Mifflin Harcourt.

Theise, N. (2023). *Notes on complexity: A scientific theory of connection, consciousness, and being.* Spiegel & Grau.

Wilczek, F. (2021). *Fundamentals: Ten keys to reality.* Penguin Press.

Design never ends